IMPROVING

RISK

COMMUNICATION

Committee on Risk Perception and Communication

Commission on Behavioral and Social Sciences and Education
Commission on Physical Sciences, Mathematics, and Resources

National Research Council

NATIONAL ACADEMY PRESS
Washington, D.C. 1989

National Academy Press • 2101 Constitution Avenue, N.W. • Washington, D. C. 20418

The study reported here was supported by the Agency for Toxic Substances Disease Registry, Allied-Signal, American Cyanamid Company, American Industrial Health Council, American Petroleum Institute, Bristol-Myers Company, U.S. Department of Defense, U.S. Department of Energy, Dow Chemical USA, Electric Power Research Institute, U.S. Environmental Protection Agency, Exxon Corporation, Hercules Incorporated, ILSI Risk Science Institute, Mobil Oil Corporation, Monsanto Company, Motor Vehicle Manufacturers Association, and National Science Foundation. It also has received support from the National Research Council Fund, a pool of private, discretionary, nonfederal funds that is used to support a program of Academy-initiated studies of national issues in which science and technology figure significantly. The NRC Fund consists of contributions from a consortium of private foundations including the Carnegie Corporation of New York, the Charles E. Culpeper Foundation, the William and Flora Hewlett Foundation, the John D. and Catherine T. MacArthur Foundation, the Andrew W. Mellon Foundation, the Rockefeller Foundation, and the Alfred P. Sloan Foundation; and the Academy Industry Program, which seeks annual contributions from companies that are concerned with the health of U.S. science and technology and with public policy issues with technological content.

Library of Congress Cataloging-in-Publication Data

Improving Risk Communication / Committee on Risk Perception and Communication, Commission on Physical Sciences, Mathematics, and Resources, Commission on Behavioral and Social Sciences and Education, National Research Council.

    p. cm.

    Bibliography: p.

    Includes index.

    ISBN 0-309-03946-0. ISBN 0-309-03943-6 (pbk.).

    1. Risk Communication. I. National Research Council (U.S.). Committee on Risk Perception and Communication.

T10.68.I47 1989   89-9464

363.1–dc20   CIP

Printed in the United States of America

First Printing, September 1989
Second Printing, January 1990

COMMITTEE ON RISK PERCEPTION
AND COMMUNICATION

JOHN F. AHEARNE, Vice President, Resources for the Future, Washington, D.C., *Chairman*

ERNESTA BALLARD, Consultant, Seattle, Washington

RUTH FADEN, Professor, Department of Health Policy and Management, Johns Hopkins University, Baltimore, Maryland

JAMES A. FAY, Professor, Department of Mechanical Engineering, Massachusetts Institute of Technology, Cambridge

BARUCH FISCHHOFF, Professor, Department of Engineering and Public Policy and Department of Social and Decision Sciences, Carnegie-Mellon University, Pittsburgh, Pennsylvania

THOMAS P. GRUMBLY, President, Clean Sites, Inc., Alexandria, Virginia

PETER BARTON HUTT, Partner, Covington & Burling, Washington, D.C.

BRUCE W. KARRH, Vice President, Safety, Health & Environmental Affairs, E. I. du Pont de Nemours & Co., Wilmington, Delaware

D. WARNER NORTH, Principal, Decision Focus, Inc., Los Altos, California; Consulting Professor, Department of Engineering-Economic Systems, and Associate Director, Center for Risk Analysis, Stanford University, Stanford, California

JOANN E. RODGERS, Deputy Director of Public Affairs and Director of Media Relations, The Johns Hopkins Medical Institutions, Baltimore, Maryland

MILTON RUSSELL, Professor of Economics, University of Tennessee, Knoxville, and Senior Economist, Oak Ridge National Laboratory, Oak Ridge, Tennessee

ROBERT SANGEORGE, Vice President for Public Affairs, National Audubon Society, New York

HARVEY M. SAPOLSKY, Professor of Public Policy and Organization, Political Science Department, Massachusetts Institute of Technology, Cambridge

JURGEN SCHMANDT, Professor, LBJ School of Public Affairs, University of Texas, Austin, and Director, Center for Growth Studies, Houston Area Research Center, The Woodlands, Texas

MICHAEL SCHUDSON, Chair, Department of Communication, and Professor, Department of Sociology, University of California at San Diego, La Jolla

KARL K. TUREKIAN, Department of Geology and Geophysics, Yale University, New Haven, Connecticut

RAPHAEL G. KASPER, *Executive Director*
LAWRENCE E. MCCRAY, *Associate Executive Director* (until August 1, 1988)
MYRON UMAN, *Associate Executive Director* (as of August 1, 1988)

The National Academy of Sciences is a private, nonprofit, self-perpetuating society of distinguished scholars engaged in scientific and engineering research, dedicated to the furtherance of science and technology and to their use for the general welfare. Upon the authority of the charter granted to it by the Congress in 1863, the Academy has a mandate that requires it to advise the federal government on scientific and technical matters. Dr. Frank Press is president of the National Academy of Sciences.

The National Academy of Engineering was established in 1964, under the charter of the National Academy of Sciences, as a parallel organization of outstanding engineers. It is autonomous in its administration and in the selection of its members, sharing with the National Academy of Sciences the responsibility for advising the federal government. The National Academy of Engineering also sponsors engineering programs aimed at meeting national needs, encourages education and research, and recognizes the superior achievements of engineers. Dr. Robert M. White is president of the National Academy of Engineering.

The Institute of Medicine was established in 1970 by the National Academy of Sciences to secure the services of eminent members of appropriate professions in the examination of policy matters pertaining to the health of the public. The Institute acts under the responsibility given to the National Academy of Sciences by its congressional charter to be an adviser to the federal government and, upon its own initiative, to identify issues of medical care, research, and education. Dr. Samuel O. Thier is president of the Institute of Medicine.

The National Research Council was established by the National Academy of Sciences in 1916 to associate the broad community of science and technology with the Academy's purposes of furthering knowledge and of advising the federal government. The Council operates in accordance with general policies determined by the Academy under the authority of its congressional charter of 1863, which establishes the Academy as a private, nonprofit, self-governing membership corporation. The Council has become the principal operating agency of both the National Academy of Sciences and the National Academy of Engineering in the conduct of their services to the government, the public, and the scientific and engineering communities. It is administered jointly by both Academies and the Institute of Medicine. The National Academy of Engineering and the Institute of Medicine were established in 1964 and 1970, respectively, under the charter of the National Academy of Sciences.

Preface

In 1983 the National Research Council completed a study on managing risk, leading to a report *Risk Assessment in the Federal Government: Managing the Process*. This report focused on improving risk assessment and risk decisions within the government. However, a major element in risk management in a democratic society is communication about risk. Growing concern that risk communication was becoming a major problem led to the chartering of a National Research Council committee to examine possibilities for improving social and personal choices on technological issues by improving risk communication.

The National Research Council initiated the study out of recognition that technological issues, in addition to being critically important, are complex, difficult, and laden with political controversy. Because the issues are scientific and technical in content, and cut across the concerns of many government agencies, scientific disciplines, and sectors of society, the National Research Council seemed to provide an ideal forum for the conduct of such a study. Moreover, in past work on policy in the areas of risk assessment and risk management (notably, the above-mentioned report on risk assessment), the National Research Council has helped develop concepts widely used in thinking about the policy issues.

It became evident in discussions with representatives of some key federal agencies that no single agency was willing to undertake the needed study on its own or even to act as the primary source of support for a study at the National Research Council, even though representatives of several agencies recognized the importance of risk communication to their activities. As a result, the National Research Council initiated the study with its own funds, eventually receiving support for about half the cost from a consortium of federal and private sources.

To reflect the breadth of issues to be studied, the Committee on Risk Perception and Communication was made responsible to two major units of the National Research Council, the Commission on Physical Sciences, Mathematics, and Resources and the Commission on Behavioral and Social Sciences and Education. The committee represents a cross section of many relevant kinds of experience and expertise. It includes members with extensive experience analyzing, managing, and communicating about diverse risks, including those from radiation, chemicals, drugs, disease, and consumer products. Members have experience in diverse settings, including federal and local decision-making bodies, industry, the mass media, and environmental and citizens' groups. The committee also exhibits diverse disciplinary backgrounds, including physical and social sciences, law, journalism, public health, and communications research. The National Research Council has tried in constituting the committee to achieve a balance of perspectives on all these dimensions.

The committee's charge was to offer knowledge-based advice to governments, private and nonprofit sector organizations, and concerned citizens about the process of risk communication, about the content of risk messages, and about ways to improve risk communication in the service of public understanding and better-informed individual and social choice. This report does not provide a set of prescriptional guidelines, a "how-to" manual for risk communicators. The committee concluded that many participants in the process lack fundamental understanding of the important points that form the basis for successful risk communication. Therefore this report concentrates on developing those points. The committee believes that without such understanding detailed guidelines would not be useful. With such understanding, organizations should be able to develop their own guidelines to fit their own somewhat unique functions.

Committee members met six times during the period from May 1987 to June 1988. The committee sought knowledge from several sources: experimental research on processes of perception, cognition, and understanding in individuals, including studies of the understanding of risk estimates; laboratory and field research on the conditions affecting the effectiveness of communications; and the collected experience of individuals and organizations that have engaged in organized communication about risk. The committee discussed a wide range of hazards, including but by no means restricted to those posed by toxic and carcinogenic substances and by radioactivity. It considered communication both about social choices, such as whether or how strictly to regulate hazardous substances or processes, and about personal choices, such as whether to change eating habits to avoid cancer or sexual habits to avoid AIDS. And the committee considered addressing advice to several audiences, including public agencies at all levels of government; legislatures; firms and industrial associations; environmental, consumer, and citizens' groups; journalists and mass media organizations; scientists and the organizations that employ them; and the interested public.

This report presents the insights of the committee. The report should significantly improve the understanding of what the problems are in risk communication, particularly the risk communication activities of government and industry. The committee's recommendations, if followed, would significantly improve the risk communication process.

> JOHN F. AHEARNE, *Chairman*
> Committee on Risk Perception
> and Communication

Acknowledgments

Although this report represents the work of the committee, it would not have been produced without the support of professional staff from the National Research Council, who drafted the chapters and refined them on the basis of the committee's discussions and conclusions: Paul Stern (Chapters 1 through 4), Rob Coppock (Chapters 5 and 6), and Lawrence McCray (Chapter 7). Their resumes are included with those of the committee because of their intellectual contributions, which advanced the committee's efforts throughout the study. The report was greatly improved by the diligent work of its editor, Roseanne Price. In addition, invaluable support was provided by Deborah Reischman for the first half of the committee's tenure and Nancy Crowell for the second half.

The committee acknowledges with appreciation presentations made at committee meetings by the following persons:

FREDERICK W. ALLEN, Associate Director, Office of Policy Analysis, U.S. Environmental Protection Agency

BETSY ANCKER-JOHNSON, Vice President, General Motors Corporation

GERALD L. BARKDOLL, Associate Commissioner for Planning and Evaluation, U.S. Department of Health and Human Services, Food and Drug Administration

RICHARD BAXTER, Senior Vice President, The Roper Organization

DON BERRETH, Director, Office of Public Affairs, Centers for
Disease Control

D. CHRISTOPHER CATHCART, Associate Director for Health and
Safety, Chemical Manufacturers Association

JOAN CLAYBROOK, President, Public Citizen

DEVRA DAVIS, Director, Board on Environmental Studies and
Toxicology, National Research Council

ANN FISHER, Manager, Risk Communications Program, U.S.
Environmental Protection Agency

LOWELL HARMISON, Deputy Assistant Secretary for Health, U.S.
Department of Health and Human Services

NANCY HOLLAND, Executive Director, American Blood
Commission

THOMAS H. ISAACS, Deputy Associate Director, Office of Geologic
Repositories, Office of Civilian Radioactive Waste Management,
U.S. Environmental Protection Agency

EDWARD KLEIN, Director, TOSCA Assistance Office, U.S.
Environmental Protection Agency

ANTHONY Z. ROISMAN, Cohen, Milstein & Hausfeld, Washington,
D.C.

BEN C. RUSCHE, Director, Office of Civilian Radioactive Waste
Management, U.S. Environmental Protection Agency

CHRISTINE RUSSELL, Alicia Patterson Foundation Fellow (on leave
from The Washington Post) and President, National
Association of Science Writers

LINDA SMITH, former Executive Committee member, STOP IT,
Warren, Massachusetts

ROGER STRELOW, Vice President, Corporate Environmental
Programs, General Electric Company

LEE M. THOMAS, Administrator, U.S. Environmental Protection
Agency

Contents

IMPROVING

RISK

COMMUNICATION

Summary

A NEW PERSPECTIVE

Hazards of modern life surround us and so, too, does communication about the risks of those hazards. News reports describe such hazards as pollutants in the air and in drinking water, pesticide residues in food, threats from radiation and toxic chemicals, and AIDS. Government and industry also send out messages about hazards and their risks, sometimes directly to the populace but more often through intermediaries, such as the print and broadcast media.

Risk messages are difficult to formulate in ways that are accurate, clear, and not misleading. One reads, for example, that "radon risk can equal or exceed the 2% risk of death in an auto accident, . . . for anyone who lives 20 years at levels exceeding about 25 picocuries per liter" (Kerr, 1988). This statement places an unfamiliar risk (radon exposure in homes) in juxtaposition to a more familiar risk (death in an auto accident), which may help people understand the magnitude of this unfamiliar risk. But this simple comparison may be misleading because it does not specify the respective levels of exposure, leaves out potentially relevant nonlethal consequences, and uses language (picocuries per liter) unfamiliar to most people. This report addresses these and other problems confronting risk communication.

1

Risk messages can be controversial for many reasons. The hazards they describe are often themselves centers of controversy. Frequently, there is enough uncertainty in the underlying knowledge to allow different experts to draw contradictory conclusions. Experts are frequently accused of hiding their subjective preferences behind technical jargon and complex, so-called objective analyses. Often a message that is precise and accurate must be so complex that only an expert can understand it. Messages that nonexperts can understand necessarily present selected information and are thus subject to challenge as being inaccurate, incomplete, or manipulative.

In the past the term risk communication has commonly been thought of as consisting only of one-way messages from experts to nonexperts. In this report the Committee on Risk Perception and Communication takes a different perspective. Because much of the controversy seems to center on the content of specific messages, it was tempting to proceed along the lines of many previous discussions about risk communication and concentrate on message design. We found a focus on one-way messages too limiting, however. Instead, we make a crucial distinction between risk messages and the risk communication process. We see risk communication as an interactive process of exchange of information and opinion among individuals, groups, and institutions. When risk communication is viewed in this way, significant, though perhaps less obvious, underlying problems can be better discerned and treated.

We view success in risk communication in a different way also. Some take the position that risk communication is successful when recipients accept the views or arguments of the communicator. We construe risk communication to be successful to the extent that it raises the level of understanding of relevant issues or actions for those involved and satisfies them that they are adequately informed within the limits of available knowledge.

Finally, it should be noted that one of the most difficult issues we faced concerned the extent to which public officials in a democratic society should attempt to influence individuals—that is, to go beyond merely informing them—concerning risks and such risk-reducing actions as quitting smoking. Government officials must be accountable for their decisions and will likely find their efforts to influence contested if they stray from accepted scientific views or if they challenge popular consensus. A public official should be aware of the political risks and of the legitimate constraints placed upon government in advocacy. Procedural strategies such as independent review processes can be used to determine the appropriateness of the use of influencing techniques. Where an unusually strong degree of advocacy seems warranted, officials should seek legitimization of such actions through the democratic process.

COMMON MISCONCEPTIONS ABOUT RISK COMMUNICATION

Several important misconceptions need to be dispelled before the real problems of risk communication can be addressed. Contrary to what some think, there is no single overriding problem and thus no simple way of making risk communication easy. Risk messages necessarily compress technical information, which can lead to misunderstanding, confusion, and distrust.

Many people—including some scientists, decision makers, and members of the public—have unrealistic expectations about what can be accomplished by risk communication. For example, it is mistaken to expect improved risk communication to always reduce conflict and smooth risk management. Risk management decisions that benefit some citizens can harm others. In addition, people do not all share common interests and values, so better understanding may not lead to consensus about controversial issues or to uniform personal behavior. But even though good risk communication cannot always be expected to improve a situation, poor risk communication will nearly always make it worse. It is also mistaken to think, as some do, that if people understood and used risk comparisons it would be easy for them to make decisions. Comparing risks can help people comprehend the unfamiliar magnitudes associated with risks, but risk comparison alone cannot establish levels of acceptable risk or ensure systematic minimization of risk. Factors other than the level of risk—such as the voluntariness of exposure to the hazard and the degree of dread associated with the consequences—must be considered in determining the acceptability of risk associated with a particular activity or phenomenon.

Some risk communication problems derive from mistaken beliefs about scientific research on the nature of how risks are assessed and managed and on risk communication itself. Scientific information, for example, cannot be expected to resolve all important risk issues. All too often research that would answer the question has not been done or the results are disputed. Although a great deal of research has been done on the dissemination and preparation of risk messages, there has been much less attention devoted to the risk communication process. In addition, even when valid scientific data are available, experts are unlikely to agree completely about the meaning of these data for risk management decisions. Finally, it is unrealistic to expect easy identification and understanding of the values, preferences, and information needs of the intended recipients of risk messages.

Other misconceptions involve stereotypes about the way intermediaries and recipients react to risk messages. It is mistaken, for example, to view journalists and the media always as significant, independent causes of problems in risk communication. Rather, the problem is often at the interface between science and journalism. Both sides need to better understand the pressures and constraints of the other instead of complaining about the sometimes disappointing results. Scientists and risk managers should recognize the importance of the part journalists play in identifying disputes and maintaining the flow of information during resolution of conflicts; journalists need to understand how to frame the technical and social dimensions of risk issues. It is also important to recognize the differences between the broadcast and the print media and between the national and the regional or local press corps.

Finally, even though most people prefer simplicity to complexity, it is mistaken to expect the public to want simple, cut-and-dried answers as to what to do in every case. The public is not homogeneous. People differ in the degree to which they exercise control over exposure to hazards or remediation of undesirable consequences, the importance they attach to various consequences, and their tendency to be risk averse or risk seeking. Often at least part of the public seeks considerable information about the risks they face.

PROBLEMS OF RISK COMMUNICATION

We distinguish two major types of problems in risk communication. Problems deriving from institutional and political systems are problems for which little can be done—beyond trying to understand them—by those involved in risk communication. Nevertheless, these problems can have a considerable impact on actions and events. Problems of risk communicators and recipients can be addressed more directly and are therefore more amenable to improvement or solution.

Problems Deriving from the Institutional and Political Systems

Several kinds of legal considerations, including statutory mandates, liability, and informed consent and "right-to-know" requirements, influence the options available to risk managers and thus the content of their risk messages. These considerations generally either limit the possible responses to the risk in question or require that certain actions be taken in given circumstances. For example, sometimes statutes require consideration of certain factors (the Federal Insecticide, Fungicide, and Rodenticide Act explicitly includes consideration of economic benefits) or the exclusion of others (the Clean Water Act specifies that the best available technology should be used regardless of the financial burden imposed). Although not necessarily problems as such, these considerations often constitute important influences on risk messages and risk communication processes. It is often difficult to understand why risk messages appear as they do without consideration of these factors.

Communicating with citizens about risks can increase their desire to participate in or otherwise influence decisions about the control of those risks, thereby making risk management even more cumbersome. The interests of citizens and their motivation to participate in the political process can introduce difficult challenges when the implementation of risk control measures is necessarily decentralized and local preferences (generally to avoid exposure to a particular risk) preclude solutions in the broader interest. Many hazardous waste facilities operate under these pressures.

Divided authority, not only among Congress, the executive branch, and the courts at the federal level but also among federal, state, and local or regional jurisdictions, creates incentives for each actor to gain as much leverage as possible from the limited portion he or she controls. Such fragmentation makes communicating about risks harder because it makes government regulation and risk reduction programs more complex and makes it more difficult to determine who is responsible for the eventual outcomes.

Government and industry spend large amounts of money on research, and thus their concerns are usually well reflected in the information developed by that research. Individuals and citizens' groups do not usually have the financial resources to fund research and thus do not enjoy this sort of access to information and influence over its generation. If a group of people that a risk communicator is trying to reach feels that the system for generating information relied upon by that source does not consider the group's concerns, it may reject the information from that source as a basis for decisions about risks. It is reasonable to speculate that this may, in part, explain why it is so difficult to affect young people's attitudes and behavior about drugs and the AIDS epidemic—the information presented is based on facts that they do not consider very important in the face of their immediate concerns of peer pressure and personal image.

There also may be systematic biases in the provision of information. Those most strongly motivated to communicate about risk are often also those with the strongest interest in the decision. Whenever a personal or social decision affects interested groups or organizations, conflicting messages reflecting the interests of those groups or organizations may be expected. The U.S. Environmental Protection Agency administrator's statement in 1984 that EDB (ethylene dibromide) contamination was a long-term health problem being adequately handled by tolerance guidelines, for example, was in the news at about the same time that public health officials in Massachusetts and Florida were removing grain products with EDB contamination from grocery store shelves. Experts from the food industry joined in, downplaying the risks, while scientists from environmental groups criticized the government's inaction. The beliefs, predispositions, and interests of risk communicators and the groups they represent create incentives to slant, or even distort or misrepresent, information. This can skew messages in many different directions on the same issue.

Problems of Risk Communicators and Recipients

The problems encountered by the sources and recipients of risk messages center on the following topics: establishing and recognizing credibility, making the messages understandable, preparing messages in an emergency, capturing and focusing attention, and getting information.

Lack of credibility alters the communication process by adding distrust and acrimony. The most important factors affecting the credibility of a source and its messages relate to the accuracy of the messages and the legitimacy of the process by which the contents were determined, as perceived by the recipients. Recipients' views about the accuracy of a message are adversely affected by (1) real or perceived advocacy by the source of a position in the message that is not consistent with a careful assessment of the facts; (2)

a reputation for deceit, misrepresentation, or coercion on the part of the source; (3) previous statements or positions taken by the source that do not support the current message; (4) self-serving framing of information in the message; (5) contradictory messages from other credible sources; and (6) actual or perceived professional incompetence or impropriety on the part of the source. The perceived legitimacy of the process by which the contents of a message were determined depends on (1) the legal standing of the source with respect to the risks addressed; (2) the justification provided for the communication program; (3) the access afforded affected parties to the decision-making process; and (4) the degree to which conflicting claims are given fair and balanced review.

Ideally, risk information should use language and concepts recipients already understand. It is difficult to present scientific and technical information that uses everyday language and magnitudes common in ordinary experience and that is sensitive to such psychological needs on the part of recipients as the desire for clear, decisive answers or the fear of the unfamiliar and unknown.

Sometimes risk communicators must disseminate messages when there are not enough relevant data to allow them to draw satisfactory conclusions and there is no time to obtain better information. This usually occurs when an emergency requires that action be taken immediately or not at all or when events lead to requests for information prior to the completion of study or analysis.

Many things compete with risk messages for attention, and it is often difficult to get the intended recipients to attend to the issues the risk communicator thinks are important. From the risk communicator's standpoint, there are two aspects of this: stimulating the attention of the ultimate recipient and interacting with the news media and other intermediaries. There are, of course, several different ways that messages can reach the final recipients: face-to-face (physician to patient, friend to friend, within the family), in groups (work sites, classrooms), through professional or volunteer organizations (American Medical Association, Red Cross), through the mass media (radio, television, magazines, newspapers, direct mail, billboards), and through community service agencies (at libraries, hospitals, malls, fairs).

Recipients of risk messages may have difficulty deciding which issues to attend to or what to do because they cannot get information from officials and other message sources that satisfactorily answers their questions. This can happen when authorities do not listen and therefore do not provide what the recipient considers relevant information or because the individual is unable to find a trusted source or interpreter of already available information.

CONCLUSIONS AND RECOMMENDATIONS

In formulating recommendations we focused on the preparation and dissemination of formal risk messages to audiences that include nonexperts and on only two of the many types of risk-managing organizations: government agencies and large private corporations. Nevertheless, our recommendations are intended to attack the problems of recipients of risk messages as well. The goal cannot be only to make those who disseminate formal risk messages more effective by improving their credibility, understandability, and so on. Such an approach might serve their interests, but it could well degrade the overall quality of risk communication if it merely meant that they could advance their viewpoints with greater influence. Risk communication can be improved only if recipients are also helped to solve their problems at the same time.

The risk communication process—usually with many messages from many sources—can be considered successful only to the extent that it, first, improves or increases the base of accurate information that decision makers use, be they government officials, industry managers, or individual citizens, and, second, satisfies those involved that they are adequately informed within the limits of available knowledge. This does not always result in the responses a particular source might wish, nor does it always lead to consensus about controversial issues or to uniform personal behavior. People do not all share common interests and values, and so better understanding will not necessarily lead them all to the same conclusion.

Improving risk communication is therefore more than merely crafting "better messages." Risk communication procedures as well as risk message content must be improved. Because risk communication is so tightly linked to the management of risks, solutions to the problems of risk communication often entail changes in risk management and risk analysis. Once the constraints, limitations, and incentives affecting the preparation and dissemination of messages—as well as how these factors become manifest in what we call the risk communication process—are understood, improvements can be implemented.

This is not to imply, however, that there is a single shortcut to improving the nation's risk communication efforts. The needed improvement can come only incrementally and only from careful attention to many details. Risk managers need to consider risk communication as an important and integral aspect of risk management.

Four sets of recommendations are presented: (1) recommendations that pertain to the processes that source organizations use to generate decisions, knowledge, and risk messages; (2) recommendations that pertain to the content of individual risk messages; (3) a call for a "consumer's guide" that will enhance the ability of other groups or individuals to understand and participate in risk management activities; and (4) a brief summary of research needs.

Two broad themes run through the process and content recommendations. The first is the recognition that risk communication efforts should be more systematically oriented to the intended audiences. The most effective risk messages are those that quite self-consciously address the audiences' perspectives and concerns. The second is that openness is the surest policy. A central premise of democratic government—the existence of an informed electorate—implies a free flow of information. Suppression of relevant information is not only wrong but also, over the longer term, usually ineffective.

Management of the Process

We identified four process objectives that are key elements in improving risk communication: (1) goal setting, (2) openness, (3) balance, and (4) competence.

Setting Realistic Goals

Risk communication activities ought to be matters of conscious design. Practical goals should be established that explicitly accommodate the political/legal mandates and constraints bounding the process and the roles of the potential recipients of the organization's risk messages, on the one hand, and clearly show the contribution to improved understanding of issues and actions on the other. Explicit consideration of such factors encourages realistic expectations, clarification of motives and objectives (both within the source organization and among outside groups and individuals), and evaluation of performance.

Safeguarding Openness

Risk communication should be a two-way street. Organizations that communicate about risks should ensure effective dialogue with potentially affected outsiders. This two-way process should exhibit (1) a spirit of open exchange in a common undertaking rather than a series of "canned" briefings restricted to technical "nonemotional" issues and (2) early and sustained interchange that includes the media and other message intermediaries. Openness does not ordinarily, however, imply empowerment to determine the host organization's risk management decisions. To avoid misunderstanding, the limits of participation should be made clear from the outset.

Safeguarding Balance and Accuracy in Risk Messages

In order to help ensure that risk messages are not distorted and do not appear to be distorted, those who manage the generation of risk assessments and risk messages should (1) hold the preparers of messages accountable for detecting and reducing distortion; (2) consider review by recognized independent experts of the underlying assessment and, when feasible, the message; (3) subject draft messages, if possible, to outside preview to determine if audiences detect any overlooked distortions; and (4) prepare and release for comment a "white paper" on the risk assessment and risk reduction assessment.

Fostering Competence

Risk managers need to use procedures that incorporate two distinct types of expertise: on the risk subject matter (e.g., carcinogenic risk, occupational safety) and on risk communication. Organizations that communicate about risk should take steps to ensure that the preparation of risk messages becomes a deliberate, specialized undertaking, taking care that in the process they do not sacrifice scientific quality. Such steps include (1) deliberately considering the makeup of the intended audience and demonstrating how the choice of media and message reflects an understanding of the audience and its concerns; (2) attracting appropriate communications specialists and training technical staff in communications; (3) requiring systematic assurance that substantive risk experts within the organization have a voice in producing accurate assessments and the derivative risk message; (4) establishing a thoughtful program of evaluating the past performance of risk communication efforts; and (5) ensuring that their organizations improve their understanding of the roles of intermediaries, particularly media reporters and editors, including an understanding of the factors that make a risk story newsworthy, of the practical time and space constraints, and of the limited technical background of most media personnel.

Risk Communication in Crisis Conditions

The process for risk communication in crisis conditions requires special care. Risk managers should ensure that (1) where there is a foreseeable potential for emergency, advance plans for communication are drafted, and (2) there is provision for coordinating among the various authorities that might be involved and, to the extent feasible, a single place where the public and the media can obtain authoritative and current information.

Content of Risk Messages

We identified four generic issues that have been the source of difficulty in the past over a broad range of risk communication efforts: (1) audience orientation, (2) uncertainty, (3) risk comparisons, and (4) completeness.

Relating the Message to the Audiences' Perspectives

Risk messages should closely reflect the perspectives, technical capacity, and concerns of the target audiences. A message should (1) emphasize information relevant to any practical actions that individuals can take; (2) be couched in clear and plain language; (3) respect the audience and its concerns; and (4) seek to inform the recipient, unless conditions clearly warrant the use of influencing techniques. One of the most difficult issues in risk communication in a democratic society is the extent to which public officials should attempt to influence individuals—that is, to go beyond merely informing them—concerning risks and such risk-reducing actions as quitting smoking.

Handling Uncertainty

Risk messages and supporting materials should not minimize the existence of uncertainty. Data gaps and areas of significant disagreement among experts should be disclosed. Some indication of the level of confidence of estimates and the significance of scientific uncertainty should be conveyed.

Comparing Risks

Risk comparisons can be helpful, but they should be presented with caution. Comparison must be seen as only one of several inputs to risk decisions, not as the primary determinant. There are proven pitfalls when risks of diverse character are compared, especially when the intent of the comparison can be seen as that of minimizing a risk (by equating it to a seemingly trivial one). More useful are comparisons of risks that help convey the magnitude of a particular risk estimate, that occur in the same decision context (e.g., risks from flying and driving to a given destination), and that have a similar outcome. Multiple comparisons may avoid some of the worst pitfalls. More work needs to be done to develop constructive and helpful forms of risk comparison.

Ensuring Completeness

A complete information base would contain five types of information: (1) on the nature of the risk, (2) on the nature of the benefits that might be changed if risk were reduced, (3) on the available alternatives, (4) on uncertainty in knowledge about risks and benefits, and (5) on management issues. There are major advantages in putting the information base into written form as an adjunct to the risk message.

A Consumer's Guide to Risk and Risk Communication

Major government and private organizations that sustain risk communication efforts should jointly fund the development of a Consumer's Guide to Risk and Risk Communication. The purpose of this guide would be to articulate key terms, concepts, and trade-offs in risk communication and risk management for the lay audience, to help audiences discern misleading and incomplete information, and to facilitate the needed general participation in risk issues. Such a guide should (1) involve support from, but not control by, the federal government and other sources of risk messages; (2) be under the editorial control of a group that is clearly oriented toward the recipients of risk messages and under administrative management by an organization that is known for its independence and familiarity with lay perspectives and that can undertake the needed outreach and public information effort; and (3) cover subjects such as the nature of risk communication, concepts of zero risk and comparative risk, evaluation of risk messages, and others designated by project participants.

Research Needs

As a result of our deliberations, we have identified nine research topics for attention: (1) risk comparison, (2) risk characterization, (3) role of message intermediaries, (4) pertinency and sufficiency of risk information, (5) psychological stress, (6) the "mental models" of recipients, (7) risk literacy, (8) retrospective case studies of risk communication, and (9) contemporaneous assessment of risk management and risk communication. Two criteria guided their selection: (1) that additional knowledge would lead to material improvement in risk communication practices and (2) that creation of such knowledge is likely given past results and current research methods. We have not assigned priorities among the nine topics.

1
Introduction

When government agencies must decide whether to evacuate people from areas where toxic substances are leaching from waste dumps, set standards for exposure to suspected carcinogens, or decide whether to license nuclear power plants despite some low probability of rupture in a future earthquake, democratic societies are faced with difficult choices. The usual criteria of consensus or social acceptability are insufficient to resolve such issues of modern technology. The decisions also need to be scientifically informed, because some choices set in motion physical or biological processes whose results, if they could be foreseen, would be considered undesirable by most people.

Only a few experts possess the best knowledge available to estimate accurately the extent of the possible harm or the likelihood of its occurrence. But while great weight needs to be given to the specialized knowledge of these experts, democratic principles require that the decisions be controlled by officials, generally nonspecialists, who are answerable to the public. As Jefferson realized long ago, public decisions that require specialized knowledge raise questions about political power.

> I know of no safe depository of the ultimate powers of society but the people themselves; and if we think them not enlightened enough to exercise their control with a wholesome discretion, the remedy is

14

not to take it from them, but to inform their discretion. (Thomas Jefferson, letter to William Charles Jarvis, September 28, 1820)

To remain democratic, a society must find ways to put specialized knowledge into the service of public choice and keep it from becoming the basis of power for an elite.

Because technological decisions have implications for public health and for political power, they are often highly contentious and emotional. Participants, expert and nonexpert alike, have much at stake and are strongly motivated to work for the outcomes they favor. The ensuing political struggles are often frustrating for the participants. Nonexperts are frustrated by the inaccessibility of the knowledge they need to inform their opinions and by presentations of needed knowledge that are oversimplified, overly technical, or condescending in tone. Technical experts are frustrated when their explanations of available knowledge are met with apathy, disbelief, or anger. Government and corporate officials are frustrated when their discussions of technological alternatives are met by expressions of public mistrust and accusations of malevolence. Environmental activists are frustrated by requirements that they argue positions that are based on human and environmental values in the language of science and technology and by lack of sufficient resources to make technical arguments strongly.

Participants come to see the debates in different ways, depending on their positions in them and the frustrations they have experienced (Dietz et al., 1989; Edwards and von Winterfeldt, 1986; Lynn, 1986; Otway and von Winterfeldt, 1982). Many, especially in the scientific and technical community and in government, have defined the underlying problem in terms of "public understanding of risk," "risk perception," and "risk communication." They believe that what is needed is for people to better understand or more accurately perceive the potential costs and benefits of certain technological options. To accomplish this, they argue that scientists, governments, and the mass media need to do a better job of risk communication, by which they mean explaining the choices and their likely consequences to nonexperts. They argue that increased efforts of this kind would make conflicts about technological choices easier to resolve and would enable the society to make better choices about protecting public health, safety, and environmental quality. For reasons elaborated throughout this report, we believe this concept of risk communication and decision making is incomplete and, in important respects, misleading; it supports misconceptions about the risk communication process and raises unrealistic expectations about what risk communication can accomplish.

THE NEW INTEREST IN "RISK COMMUNICATION"

Interest in "risk communication" is quite recent.[1] That interest is evident in a recent explosion of conferences, seminars, articles, and books with the term "risk communication" in their titles (Bean, 1987; Covello et al., 1987b, 1988; Davies et al., 1987; Fischhoff, 1987; Lind, 1988; Otway, 1987; Plough and Krimsky, 1987; Zimmerman, 1987). It reflects increased attention, especially in some agencies of the federal government but in other organizations as well, to the task of informing the general public about the nature of the health, safety, and environmental risks associated with personal and societal choices. The new concern with informing the public has several motivating sources, not entirely consistent with each other, including (1) a requirement for or desire by government to inform, (2) a desire by government or industry officials to overcome opposition to decisions, (3) a desire to share power between government and public groups, and (4) a desire to develop effective alternatives to direct regulatory control. Moreover, the term risk communication has different meanings to different users.

Requirement for or Desire by Government to Inform

Sometimes government is required to inform the public. A series of federal laws, beginning with the Administrative Procedures Act of 1946 and continuing with the Freedom of Information Act, the National Environmental Policy Act, and the "Community Right to Know" provisions of Title III of the Superfund Amendments and Reauthorization Act of 1986, recognizes the public's right to be informed about certain hazards and risks, even if they have no part in the decision-making process. Many of these laws have been reinforced by federal court decisions and presidential executive orders. These actions emphasize the government's responsibility to be accountable to the people; they state as national policy that regulators must explain why one course was chosen rather than another and that the public has a right to see and challenge the basis for the decisions. Thus agencies are required to send messages to the public about the reasons for their decisions and to solicit messages of comment from citizens. The term risk communication is sometimes used to describe these messages.

Some government officials provide information not required by law. Regulatory officials sometimes do this because they believe that people would benefit from specific knowledge. For instance, the U.S. Environmental Protection Agency (EPA) over the past few years has made an effort to inform householders about the hazard of radiation exposure from indoor radon. And public health officials, responding to their general mandate, have long offered information to citizens about the health risks of dietary and sexual habits, drug and alcohol use, and other personal activities. Provision of such information is what some public health officials mean by risk communication.

Desire to Overcome Opposition to Decisions

Over the past 30 years public participation in debates on technological issues has intensified. More groups have become involved, including workers potentially at risk from hazardous activities, regulatory organizations, citizens' and environmental groups, the press, and the courts. The proponents of controversial technological options or decisions, most frequently in government or industry, often meet intense political opposition. Frequently, groups of citizens who oppose particular technological projects have delayed or stymied those projects with lawsuits, mobilization of congressional opposition, or public demonstrations. When a government or industry official has the benefit of extensive scientific study and the opposition seems simply to disregard the technical evidence, the official can come to see "the public" as irrational. Government and industry officials who see the issues this way are likely to define the conflicts that surround them as debates between the informed and the ignorant or, worse, between the rational and the irrational. Such officials are tempted to look for ways to influence the members of the opposition, either by more actively presenting a straightforward account of the knowledge they have available or by carefully packaging or even distorting that knowledge to achieve a persuasive effect. The use of information to overcome political opposition makes some notion of risk communication attractive to many proponents of controversial technology; it is, in fact, what they mean by the term.

Desire to Share Power Between Government and Public Groups

Government officials have sometimes seen in risk communication a way to reduce conflict with segments of the public by sharing power. In this situation an agency takes the role of technical analyst and

adviser, gathering and summarizing the information relevant to a
decision at hand and explaining that information to various political
actors. The agency's role might be to inform a public debate, for
instance, in a legislative decision on siting of a hazardous facility.
Or, if the agency is legally required to make the decision itself, it
can provide information to the public and use the ensuing debate to
help arrive at a decision that it judges to be both defensible within
its legal mandate and maximally acceptable to the interested parties
involved.

Such an approach was used, albeit unsuccessfully, by the EPA in
a controversial case in 1983. Prior to making a regulatory decision
about an Asarco Corporation smelter that was releasing arsenic into
the air, the agency presented the people of Tacoma, Washington,
with the best information it had available about the risks and benefits of three possible outcomes: continued operation of the smelter,
operation with pollution controls added, and closing of the smelter
(Krimsky and Plough, 1988). EPA intended that the ensuing dialogue would help the community arrive at its own preference and
inform EPA so it could make a defensible decision that would also
satisfy local opinion. Administrator William Ruckelshaus justified
his action, which depended critically on the success of the agency's
efforts to provide information, with an appeal to Jefferson's advice to
inform the public's discretion (Ruckelshaus, 1983). The incident led
to a heated controversy in which EPA was accused by some of an eva-
sion of its responsibility and by others of attempting to manipulate
the public by presenting an incomplete set of options. Although the
smelter was shut down by Asarco before the public process ran its
course, Ruckelshaus's goal of achieving consensus appears unlikely
to have been attained.

Desire to Develop Effective Alternatives to Direct
Regulatory Control

Government officials sometimes wish to persuade individuals to
protect their health by personal action rather than to adopt regulatory policies that require health-protective actions. The new interest
in risk communication in government partly reflects a search for alternatives to direct regulatory control of health hazards, which was
accelerated in the 1980s by the Reagan administration's philosophical opposition to regulation. Government agencies have sought ways
to control hazardous substances or activities short of banning them

(as they have done with high-dose vitamin preparations), restricting or taxing them (as with alcohol), or requiring control measures (as with seat belts). Some of the alternatives involve replacing regulatory prohibitions and financial penalties imposed on those who produce hazardous technologies with reliance on informed discretion of the users of those technologies. For instance, the early 1980s brought a shift in the government's treatment of most motorists' unwillingness to use seat belts. A regulatory requirement that manufacturers equip cars with air bags or other "passive restraints" to protect passengers who fail to fasten seat belts was replaced by a campaign to persuade, relying on paid and public service advertising. The government even supported research on better ways to convince people to use seat belts (Geller, 1983). Now, after many years, there appears to be an increase in the use of seat belts, although in some cases this may be due to state laws mandating their use. Such persuasion programs are adequate as alternatives to regulatory constraint only if two conditions are met: if persuasion is accepted as a technique of public policy and if persuasion is about as effective as direct regulatory control. To some an important aspect of risk communication is the use of messages to induce people to protect themselves.

A NEW DEFINITION OF RISK COMMUNICATION

Although the motives for and meanings of risk communication described above are very different in some ways, each emphasizes a particular kind of message: a message that is developed by technical experts; that describes or characterizes hazards, risks, or risk-reducing actions; and that is addressed to nonexperts. To many who use the term, risk communication means simply the development and delivery of this kind of one-way message. This widespread usage is illustrated in the foreword to the published proceedings of the first National Conference on Risk Communication, attended by 500 people in Washington, D.C., in 1986. William Reilly, the president of the Conservation Foundation, observed:

> [In] the conflict or confusion over risk questions . . . often the communication process is at fault or, at the least, exacerbates the problem. Risk communicators simply do not do a good job of getting their message across. (Reilly, 1987:vii)

This very typical formulation equates risk communication with the delivery of certain kinds of messages—one-way messages from government or other risk communicators to the general public about the nature of risks. It defines the success of risk communication from the point of view of the senders of those messages, in terms of "getting the message across." The image is of experts enlightening or persuading an uninformed and passive public.

We consider this formulation of the problem to be incomplete in critical ways. Increased efforts "to get the message across" by describing the magnitude and balance of the attendant costs and benefits or by telling people which option provides the greatest net benefit to society will have little effect for several reasons. First, costs and benefits are not equally distributed across a society. Those who bear more than a proportionate share of the costs of one of the options want to convince others that the selection of that alternative would be unfair to them. Other political participants want to make similar arguments on their own behalf or to consider the arguments of all the interested parties. Thus an important aspect of conflicts about technological issues is that these are often conflicts between different interest groups. These conflicts cannot be resolved simply by knowledge about the likely effects of each alternative on the society as a whole or on various groups.

Second, people do not agree about which harms are most worth avoiding or which benefits are most worth seeking. They want to argue for the protection of what they value and to consider which values are most worth preserving or advancing in each decision context. Because conflicts about technological issues pit values against each other, it is impossible to calculate net benefit to society—or even to subgroups of the society—on any scale that will satisfy all the participants. Values need to be debated and weighed in a political process.

Third, citizens of a democracy expect to participate in debate about controversial political issues and about the institutional mechanisms to which they sometimes delegate decision-making power. A problem formulation that appears to substitute technical analysis for political debate, or to disenfranchise people who lack technical training, or to treat technical analysis as more important to decision making than the clash of values and interests is bound to elicit resentment from a democratic citizenry. Because of such reactions to them, problem formulations that attribute technological conflict to widespread public ignorance only exacerbate the conflict.

We do not deny or minimize the importance of scientific and technological knowledge to informed public decisions about technology. In fact, we strongly endorse the proposition that understanding science in general and the likely consequences of particular technological choices should be more widespread. But we emphasize just as strongly the fact that technological choices are value laden. Nonexperts need to gain technical knowledge, but technical experts and public officials also need to learn more about nonexperts' interests, values, and concerns.

In a democracy communication is an essential part of all societal decisions. The participants—individuals, groups, and institutions—express their concerns and viewpoints, present facts and arguments to support them, and listen to what other participants have to say. At various points in this ongoing process, elected officials and public servants act in the name of the society, sometimes adding their own messages to those already current. The communication continues, with concerns and viewpoints about government actions and messages as well as about the original issues being addressed.

We see risk communication as a particular instance of this sort of democratic dialogue. Thus we have come to use the term risk communication differently from its common current usage.[2] *Risk communication is an interactive process of exchange of information and opinion among individuals, groups, and institutions. It involves multiple messages about the nature of risk and other messages, not strictly about risk, that express concerns, opinions, or reactions to risk messages or to legal and institutional arrangements for risk management.* As we will establish in Chapter 4, risk communication is successful only to the extent that it raises the level of understanding of relevant issues or actions and satisfies those involved that they are adequately informed within the limits of available knowledge.

Risk communication is a component of risk management, which is the selection of risk control options. It is the process that provides the information on which government, industry, or individual decision makers base their choices. Successful risk communication does not guarantee that risk management decisions will maximize general welfare; it only ensures that decision makers will understand what is known about the implications for welfare of the available options.

The above definition of risk communication differs critically from many common uses in distinguishing between communication, which is an interactive process, and messages, which flow one way. Among people responsible for designing messages about risk, there is a temptation to confuse the task of message design and dissemination with the entire communication process and to equate the success of their messages in producing the effect desired with the success of risk communication. We have chosen a definition that takes a broader perspective than that of any single participant in the process in order to emphasize the difference between the disparate activities and goals of the many participants and the social purposes of the risk communication process.

Risk communication includes all messages and interactions that bear on risk decisions. Thus risk communication includes announcements, warnings, and instructions moving from expert sources to nonexpert audiences—the kinds of messages Reilly refers to. But it also includes other kinds of messages—about risk information and information sources, about personal beliefs and feelings concerning risks and hazards, and about reactions to risk management actions and institutions. Not all these messages are strictly about risk, but all are material to risk management.

Our use of the term risk communication also pays explicit attention to the social interaction and debate that are essential to democratic political choice and that often contribute to personal decisions about hazardous activities. Risk communication includes messages moving in various directions—not only from experts to nonexperts but also from nonexperts to each other, from nonexperts to experts, and especially the messages of political participation, from citizens to public decision makers. Decisions in government depend on dialogue between the decision maker and staff within the responsible agency and between the decision maker and various political participants, who influence the decision maker's view of the risks and the risk management options. Messages about nonexperts' perceptions of fairness, legal constraints, feelings of outrage, and the mobilization of interest group pressure are among the important elements of the risk communication process, along with messages about the risks themselves. Even with personal risk decisions, choice often depends on a dialogue in which technical knowledge may not be the dominant influence. Decisions to stop smoking, for instance, have often been influenced more strongly by the expressed value preferences of the smoker's children than by experts' messages about health consequences.

As with other communication in a democracy, the intent of the participants in risk communication is sometimes political. That is, messages about risk are sometimes intended to influence the beliefs or actions of those to whom they are addressed. Risk communication, then, must be understood in the context of decision making involving hazards and risks, that is, risk management. Communication about risk deserves special attention because the highly technical nature of the subject matter makes it more difficult than communication about other controversial issues. Risk decision makers, including individuals managing personal hazards and participating in public decisions, need to seek and interpret complex technical information from scientific disciplines in which they have not been trained. They must communicate with, and to some extent rely upon, the experts who generate that information. Because the attendant choices are controversial, affecting important economic interests and strongly held values, participants in the decision process, including experts and their employers, have incentives to appeal to emotions, distort facts, and otherwise use communication to influence the ultimate choice in the directions they desire. Thus there are no participants in debates on technological issues on whom nonexperts and public officials can rely unquestioningly for unbiased information.

RISK MESSAGES AS PART OF THE
RISK COMMUNICATION PROCESS

Risk messages, because they flow in only one direction, are only part of the interactive risk communication process. Risk messages include verbal statements, pictures, advertisements, publications, legal briefs, warning signs, or other declaratory activities that describe, characterize, or advocate positions or actions regarding risks, hazardous technologies or activities, or risk control options. Each risk message has an identifiable source and is addressed to one or more audiences.

Risk messages come from a variety of sources: physicians, journalists, regulatory agencies, manufacturers, environmental groups, health officials, and various self-appointed advisers. The messages are sometimes merely descriptive of risks and scientific studies of them; at other times the messages also describe the broad context within which a specific hazard or risk is found, the developments that preceded its occurrence, comparison of it to other hazards or risks, or the presentation of information about a risk along with information about the attendant benefits and the risks and benefits of alternatives. As mentioned above and as discussed in more detail in Chapter 4, risk messages may be constructed to inform their recipients or to influence them.

A large theoretical and empirical literature on communication, social influence, and persuasion provides considerable knowledge for anyone who wants to design effective risk messages. However, this knowledge is sufficient only to identify important principles, barriers, pitfalls, opportunities, and so forth. It is not adequate to inform many of the specific choices message designers make about characterizing particular risks for particular audiences. Lessons from the communication literature have been applied with some success in a range of areas, some of which involve efforts to induce individuals to reduce risks to themselves from cigarette smoking (McAlister, 1981) and heart disease (Maccoby and Solomon, 1981).[3] The following, necessarily brief, summary gives some idea of the concepts and general conclusions developed in this research tradition. Researchers typically discuss the message content, the source of the message, the channel by which the message is transmitted, and the audience or recipients of the message (Hovland et al., 1953; McGuire, 1985).

Of key importance to the effect of a message are the characteristics of the intended audience. The important attributes of the audience include its makeup in terms of cultural background, shared interests, concerns and fears, social attitudes, and its facility with language. A message that has a desired effect on one audience may have little effect on another. Messages in scientific language are likely to mean little to nonscientists, whereas messages about risk in everyday language may be unimpressive to scientists.

Risk messages can be carried by a variety of media: face-to-face interaction, direct mailings, advertising, hot lines, presentations to groups, press conferences, television or radio interviews, newspaper or journal articles, and so on. Each medium has its advantages and limitations—for example, television reaches many people but needs visual material and is typically presented in short segments, and newspapers rely on the written word and can present longer, more complex messages but are less vivid and immediate in emotional impact. In general, the characteristics of each channel affect the type of message that can be effectively transmitted.

The characteristics of the source of a message often affect the way audiences respond to it. Among the key factors influencing the way recipients judge a message are the degree of expertise the recipients believe the source to possess and the degree of trust the recipients have in the source. The term "credibility" is used by researchers in this field to refer to an attribute of a source that derives from a combination of expertise and trust, as seen by the audience. It is possible for a source to be credible to some recipients and not others or on some issues but not others (McGuire, 1985). Thus a locally respected old farmer may be credible to neighboring farmers as a source of information on pesticide risks but may not be credible to the officials who convene a regulatory hearing. Similarly, the scientific representative of a hazardous waste disposal company may be credible to a federal regulator but not to the neighbors of a proposed waste site. The officials do not credit the farmer because of lack of technical expertise; the neighbors do not credit the company's scientist because of lack of trust.

Where there is widespread mistrust of public sources of information, people often rely on word-of-mouth or other local sources, even if their informants are less expert than those available through public sources. Because of this practice, public agencies can sometimes be more effective in delivering technical information to individual citizens by using trusted sources as intermediaries than by designing and disseminating messages themselves (Stern and Aronson, 1984). Public officials can also listen to trusted intermediaries to learn if tasks might be delegated or to save the time and expense of questionnaires or other analyses when less detail is sufficient.

Risk messages are often designed to inform nonspecialists. Because such messages involve complex and difficult concepts, presenting clear and understandable information is a tremendous challenge for message designers. The source's choice of message content depends on what it believes the audience needs to know, on what recipients can be expected to understand, and on the action or response the source hopes to engender.

Some risk messages are intended to influence the recipients' beliefs or actions. Messages are more effective at producing behavior change when, in addition to producing understanding, they are specific about any desired response and proximate in time and place to that response. Generally, single messages can be expected to have little effect on recipients' behavior, but organized programs of messages, in which different messages are aimed at different specific purposes, can be effective.

As discussed above, considerable research has been devoted to the study of messages to change individual behavior, and the resulting knowledge can help in designing more effective risk messages. But less is known about other aspects of risk communication. For instance, there has been little systematic study of ways to design more effective messages to express citizens' concerns to government or to influence the actions of organizations such as corporations or government agencies.

SUCCESSFUL RISK COMMUNICATION

A focus on risk communication implies a standpoint outside the process. It puts no particular actor or message source at the center. In this respect an emphasis on risk communication is different from one on risk messages. The source of a risk message is likely to define and assess the success of its messages according to its own criteria. It may choose to consider its messages successful when the recipients understand them, or when they believe them to be accurate, or when they do what the sender wants to influence them to do. Obviously, different sources may have conflicting goals for their risk messages. This is one difference between the success of a single source's messages and the success of the risk communication process.

It is possible to arrive at a meaningful idea of success for risk communication by considering a broad public purpose that successful risk communication serves. If a society values democratic decision making and well-informed, goal-directed individual choice, it makes sense from the societal standpoint to say that the purpose of risk communication is "to inform the discretion" of government officials, private organizations, and individuals. From that standpoint, *risk communication is successful to the extent that it raises the level of understanding of relevant issues or actions and satisfies those involved that they are adequately informed within the limits of available knowledge.*

Informed discretion for a government official is based on knowledge about the risks and benefits of the choices at hand; about the management situation, including legal and other constraints on choice; about the concerns and preferences of citizens and other political actors; and about the political environment. Corporate officials need much the same kinds of knowledge, although they need to pay special attention to the preferences of shareholders and can sometimes afford to pay less attention to the preferences of the public. Government and corporate officials usually inform themselves about risks and benefits with the help of expert employees or consultants who interpret technical knowledge for them (Chapter 2 discusses what is involved in understanding risks). They inform themselves about citizen concerns and political matters by paying attention to elected officials, the mass media, and diverse other sources. According to the above definition, the more accurate the official's understanding of those matters, the better the risk communication system.

Citizens are decision makers in their private lives and when they participate in political decisions. Thus a successful risk communication process informs their discretion, too. Citizens inform themselves by interpreting risk messages from various sources, including experts, intermediaries such as journalists, public relations officials of public agencies and corporations, and even friends and neighbors. They evaluate or balance what they know in order to reach a judgment and to make decisions regarding risks, such as whether to protest, ignore, negotiate, or take protective action. The more accurate the citizens' understanding of the issues at hand, the better the risk communication may be said to be.

Citizens are well informed with regard to personal choices if they have enough understanding to identify those courses of action in their personal lives that provide the greatest protection for what they value at the least cost in terms of those values. Citizens are well informed with regard to a public policy issue if they have enough understanding to evaluate which options provide the most protection at the least cost, both for themselves and for the things they believe the society should value. (In Chapter 4 we discuss the meaning of successful risk communication in more detail.)

It is important to make several points about the definition of successful risk communication. First, success is defined in terms of the information available to the decision makers rather than in terms of the quality of the decisions that ensue. *Successful risk communication does not always lead to better decisions because risk communication is only part of risk management.* Risk managers, including public officials and private citizens, must also take into account their public responsibilities or personal values. It is possible to understand fully what is known about the likely consequences of each available option and yet to make a "bad" choice; if this occurs, it is not because of a failure of risk communication. Consider, for instance, the head of a federal agency who is constrained by law to ban a food additive even though the risk communication process has made it clear that there are no less harmful or costly alternative additives. Making the legally required decision does not mean the communication process failed; in the long run the process may even provide impetus for changing the law. Similarly, if someone understands but disregards information about the dangers of smoking, skydiving, or riding a motorcycle without a helmet, this may not mean risk communication was at fault. People sometimes put themselves at risk with full knowledge, and observers attribute their acts to overriding values, willfulness, or addiction rather than to a failure of communication. Risk communication, even at best, can accomplish only so much.

Second, *successful risk communication need not result in consensus about controversial issues or in uniform personal behavior.* Although such objectives often serve the producers of risk messages as criteria of success for those messages, they are not appropriate criteria for the risk communication process in a democracy. To say that success requires that the recipients do or believe what a particular message source desires is to assume that that message source is a better judge of the recipients' interests than the recipients themselves. Because people do not all share common interests or values, better understanding will not lead them all in the same direction. And it will not necessarily make choices easier for decision makers in government agencies or elsewhere.

Third, according to the definition of success, the recipient must be able to achieve as complete an understanding of the information as he or she desires. In Chapter 4 we develop the reasons underlying this definition of successful risk communication. A risk communication process that disseminates accurate information is not successful unless the potential recipients achieve a sufficient understanding. Thus the risk communication process must be judged by the level of knowledge on which decision makers act rather than by the level of knowledge reflected in particular messages or even in the full set of messages accessible to decision makers. It is common for accurate messages to be ignored, misunderstood, or rejected; it is also possible for several inaccurate messages from different sources to be compared with each other in such a way as to give the recipient a fairly accurate overall picture.

Risk communication, then, is more than one-way transmission of expert knowledge to the uninformed. Certainly, messages describing expert knowledge to nonexperts play a critically important role in risk communication. They provide essential information that nonexperts cannot get from other sources. They are also essential because, by revealing expert dissent, they give nonexperts, including many government officials, an important tool for checking against omissions or excesses in any one expert's analysis. *Messages about expert knowledge are necessary to the risk communication process; they are not sufficient, however, for the process to be successful.* Thus, although experts and the organizations that disseminate their knowledge are important participants in risk communication, nonexperts also play an important active part by expressing concerns to experts and by asking or pressuring them to provide analysis of aspects of risk that they may not yet have examined in detail. They play an essential role by participating in the debate about what values ought to be applied to knowledge about risks and how they ought to be applied. Citizens' dialogue with public and industrial risk managers, even when it does not directly address risk, can be critical to risk management decisions. The broad definition of risk communication is a reminder that public decisions about risk require debate about values and interests as much as about risks because risks cannot be weighed against each other without considering values. As we will see in the next chapter, they cannot even be understood without considering values.

NOTES

1. For discussions of the recent interest in the subject of risk communication, see Plough and Krimsky (1987) and Stallen and Coppock (1987).

2. For a complete listing of the special terminology used in this report, see Appendix E.

3. Extensive reviews of the communication literature, covering well over 1000 sources, can be found in chapters of the 1985 *Handbook of Social Psychology* (McGuire, 1985; Moscovici, 1985; Roberts and Maccoby, 1985). A review of much of this literature with a focus on risk communication has been completed by Covello et al. (1987b).

2
Understanding Hazards and Risks

Throughout recorded history people have engaged in hazardous activities, and governments have taken action to control some of those activities in the public interest. But in recent times the hazards of greatest concern, and knowledge about them, have changed in ways that make informed decisions harder to reach. Once the focus was simply on the presence or absence of danger. If a food was "adulterated," if water was determined to be "impure," if a bridge or dam was declared "unsafe," or if a workplace was "dangerous," action was called for. When people called on government to take action, they wanted simple, clear-cut measures: ban sale of the food, supply pure water, condemn the bridge, eliminate the workplace hazard. But with increased understanding of the nature of the choices, it has become harder to maintain a simple view. Responsible decision makers need to know more about the alternatives than that one of them is hazardous.

In this chapter we outline the many kinds of knowledge a well-informed decision requires and the ways in which this knowledge is often incomplete and uncertain. We show how, under such conditions, the judgments of both experts and nonexperts can be affected by preexisting biases and cognitive limitations and how human values and concerns inevitably enter into the analytic process. These factors often lead experts to disagree with each other and with nonexperts about the significance of risks, even when the facts are not in dispute.

TOWARD QUANTIFICATION OF HAZARDS

One reason decision makers need more knowledge is that it has become clear that eliminating one danger can create a new one. To rid the water supply of organisms that cause typhoid and other infectious diseases, water has been chlorinated since early in this century. This action resulted in chemical reactions in the water that produced chloroform and other carcinogenic chlorinated hydrocarbons. To choose between the dangers, one must answer difficult questions: Which danger is more worth avoiding? How much decreased danger from typhoid is enough to justify a certain amount of increased danger of cancer? Experts agree that there will be fewer deaths from chlorination-induced cancer than there once were from typhoid, but is that enough information to make a decision? It may be important to consider that typhoid and cancer are very different kinds of dangers. Typhoid is an acute disease and cancer is a chronic one; typhoid is much more treatable; and there are alternatives to chlorination for preventing it, although the alternatives also present hazards, as yet poorly understood.

Society is faced with many choices that trade one danger for another and that raise similar questions. For instance, regulated commercial canning of food reduced the danger of botulism compared with home canning, but the use of lead solder in "tin" cans introduced a toxin not present in home canning jars. Lighter automobiles use less fuel and generate less air pollution, but in a collision with an older, heavier vehicle they are more dangerous to their occupants.

Societal choices also involve the benefits associated with hazards and the costs of hazard reduction. Industries that pollute air and water also provide jobs and profits; before requiring pollution controls, public officials usually want to consider the probable effects of the available options on those benefits. Cities may install traffic lights to reduce fatalities and injuries, but officials usually want to consider whether this is the best way to spend scarce revenues. Thus decision makers want good estimates of how much each alternative will reduce hazards so that they can judge the potential benefits against the potential costs.

Decision makers need detailed knowledge because it has become clear that making the world safer for most people can make it more dangerous for some. Pesticides and herbicides have helped make wholesome food more available and have helped improve the diets of low-income consumers, but they expose agricultural workers to hazardous chemicals and can be a significant polluter of water supplies. The total danger to society may have decreased greatly, but that knowledge may be of no comfort to farm workers. Nuclear power offers some people the benefit of cleaner air but may expose different people to radioactivity in the event of an accident. How is society to weigh small benefits to many against what are sometimes larger dangers for a relative few?

Decision makers need detailed knowledge for another reason as well: the hazards of greatest concern today are more difficult to observe and evaluate than the major hazards of the past. Half a century ago most of the major health and safety hazards were of immediate onset: accidents, bacterial infections, poisonings, and the like. Most of the hazards that are now controversial are of delayed onset, sometimes not being evident for decades after exposure and sometimes affecting only the offspring of those who were exposed. It can be hard to know what the hazards of a substance or activity are before a generation of experience has accumulated.

To make informed choices, it helps to look carefully and analytically at the hazards each alternative entails. It is important to develop quantitative knowledge: How much cancer might be caused by chlorinating water? How much pesticide are farm workers exposed to? For this kind of analysis, some conceptual distinctions are useful. The most basic of these is between "hazard" and "risk." *An act or phenomenon is said to pose a hazard when it has the potential to produce harm or other undesirable consequences to some person or thing.* The magnitude of the hazard is the amount of harm that may result, including the number of people or things exposed and the severity of consequence. *The concept of risk further quantifies hazards by attaching the probability of being realized to each level of potential harm.*[1] Thus an area that experiences a severe hurricane once in 200 years faces the same hazard but only one-tenth the risk of a similar area that experiences an equally severe hurricane once in 20 years. The concept of risk makes clear that hazards of the same magnitude do not always pose equal risks.

Risks of the same magnitude do not always pose equal concerns, either. Most quantitative measures of risk combine the undesirability of a hazard and its probability of occurrence into a single summary measure. Use of such summary measures can simplify large amounts of data but can be unsatisfying to people who want to consider different kinds of injuries or deaths separately because, for instance, they believe that certain types of individuals are worthy of special protection or that certain types of injuries or illnesses are especially to be avoided. Some ways of characterizing risk take such concerns into account. These involve calculating separate risk estimates for each hazardous effect, giving heavier weight to qualitative characteristics of risk (e.g., Fischhoff et al., 1984; Okrent, 1980) and using explicit measures of values and risk attitudes (Raiffa, 1968).

KNOWLEDGE NEEDED FOR RISK DECISIONS

What kinds of knowledge must be collected so that the process of communication will be an informed dialogue leading to reasonable choices? Understanding the risks is not enough, because organizations and individuals never choose between risks. Rather, they choose between options, each of which presents some risks. Each also presents benefits, which are as crucial to the choices as the risks are. Understanding risks can be difficult, but understanding the benefits of a set of decision alternatives can be as difficult. Both kinds of knowledge are needed for an informed choice.

This section outlines the many kinds of relevant knowledge. It summarizes four kinds of knowledge decision makers need: (1) about risks and benefits associated with a particular option, (2) about alternative options and their risks and benefits, (3) about the uncertainty of the relevant information, and (4) about the management situation.

Information About the Nature of Risks and Benefits

"Risk assessment" is the term generally used to refer to the characterization of the potential adverse effects of exposures to hazards. Risk assessment therefore addresses the questions listed below. "Benefit assessment," a term not commonly used, addresses many similar questions. Some benefit questions are mentioned below, in parentheses.

1. *What are the hazards of concern* as a consequence of a substance or activity? What environments, species, individuals, or organ systems might be harmed? How serious is each potential consequence? Is it reversible? (What are the benefits associated with a substance or activity? Who benefits and in what ways?)

2. *What is the probable exposure* to each hazard in total number of people or valued things? How do the exposures cumulate over time? A single exposure over a short period of time can have effects different from those due to exposure to the same amount of a hazard in several episodes or chronically at low levels over a longer period of time. (How many people benefit? How long do the benefits last?)

3. *What is the probability of each type of harm* from a given exposure to each hazard? How potent is the hazardous substance or activity at the relevant exposures? What is the relation of exposure or "dose" to response? (What is the probability that the projected benefits will actually follow from the activity in question? What events might intervene to prevent those benefits from being received? What are the probabilities of these events?)

4. *What is the distribution of exposure?* In particular, which groups receive a disproportionate share of the exposure? (Which groups get a disproportionate share of the benefits?)

5. *What are the sensitivities* of different populations of individuals to each hazard? What is the appropriate estimate of harm for highly sensitive populations that bear a significant proportion of the overall risk? What are those populations, where are they located, and what proportion of the total risk do they bear?

6. *How do exposures interact with exposures to other hazards?* Sometimes one exposure can make people more sensitive to another hazard—a synergistic effect—and, occasionally, exposure to one hazard may decrease sensitivity to another—a blocking effect. What is known about such effects?

7. *What are the qualities of the hazard?* For instance, do those exposed have an option to reduce or eliminate their exposure (and at what cost)? Would harm come to exposed people one at a time or as a mass, in a potential catastrophe? Is the hazard deadly or not? Does the harm take the form of accident or illness, acute or chronic disease, damage to the young or the old, to the living or the unborn? If the hazard is an illness, is it treatable? Is it a dread illness, such as cancer, or one that creates less of an emotional reaction? Table 2.1 lists qualities of risk that make a difference in most people's judgments. (What are the qualities of the benefits? Do they appear as increased income, saved time, physical comfort, improved health, more stable ecosystems, more beautiful surroundings, improved welfare for low-income people or the elderly, or in other forms?)

8. *What is the total population risk*, taking into account all of the above? To arrive at such an estimate, one must somehow calcu-

TABLE 2.1 Qualitative Factors Affecting Risk Perception and Evaluation

Factor	Conditions Associated with Increased Public Concern	Conditions Associated with Decreased Public Concern
Catastrophic potential	Fatalities and injuries grouped in time and space	Fatalities and injuries scattered and random
Familiarity	Unfamiliar	Familiar
Understanding	Mechanisms or process not understood	Mechanisms or process understood
Controllability (personal)	Uncontrollable	Controllable
Voluntariness of exposure	Involuntary	Voluntary
Effects on children	Children specifically at risk	Children not specifically at risk
Effects manifestation	Delayed effects	Immediate effects
Effects on future generations	Risk to future generations	No risk to future generations
Victim identity	Identifiable victims	Statistical victims
Dread	Effects dreaded	Effects not dreaded
Trust in institutions	Lack of trust in responsible institutions	Trust in responsible institutions
Media attention	Much media attention	Little media attention
Accident history	Major and sometimes minor accidents	No major or minor accidents
Equity	Inequitable distribution of risks and benefits	Equitable distribution of risks and benefits
Benefits	Unclear benefits	Clear benefits
Reversibility	Effects irreversible	Effects reversible
Origin	Caused by human actions or failures	Caused by acts of nature or God

NOTE: In selecting risks to be compared, it is helpful to keep these distinctions in mind. Risk comparisons that ignore these distinctions (e.g., comparing voluntary to involuntary risks) are likely to backfire unless appropriate qualifications are made.

SOURCE: Covello et al., 1988.

late a summation across different types of harm, people of different sensitivities, and exposures to the hazard in different amounts and in combination with various other hazards. (What is the total benefit?)

Information on Alternatives

The term "risk control assessment" may be used to describe the activity of characterizing alternative interventions to reduce or eliminate a hazard. More generally, decision makers need responses to questions such as the following about all the alternatives to any option under consideration:

1. *What are the alternatives* that would prevent the hazard in question? Some involve the choice of alternative processes or substances, while others involve action that might prevent or reduce exposure, mitigate the consequences, or compensate for damage.

2. *What are the risks of alternative actions* and of a decision not to act? How are these risks distributed? Since there are an infinite number of alternatives, it is possible to assess only a few, but a complete analysis should examine those alternatives being prominently discussed and should work to identify others worthy of consideration. (What benefits does each alternative promise, other than risk reduction?)

3. *What is the effectiveness* of each alternative? That is, how much does it reduce the risks it is intended to reduce, and how is the risk reduction distributed across relevant populations? (What benefits does each provide, and how are they distributed?)

4. *What are the costs* of each alternative, and how are these distributed across relevant populations?

Uncertainties in Knowledge About Risks and Benefits

Assessments of the risks and benefits of all available options, to be complete, should address the following questions about their own reliability:

1. *What are the weaknesses of the available data?* Information needed to estimate the risks and benefits of an activity or substance and the effects and costs of alternatives often does not exist. Sometimes experts dispute the accuracy or reliability of the data that are available. And often not enough is known to extrapolate confidently from those data to estimates of risks (or benefits) for a whole population.

2. *What are the assumptions and models* on which the estimates are based when data are missing or uncertain or when methods of estimation are in dispute? How much dispute exists among experts about the choice of assumptions and models?

3. *How sensitive are the estimates* to changes in the assumptions or models? That is, how much would the estimate change if it used different plausible assumptions about exposures or incidences of harm (or benefits) or different methods for converting available data into

estimates? What are the boundaries or confidence limits within which the correct risk (or benefit) estimate probably falls? What is the basis for concluding that the correct estimate is not likely to lie outside those bounds?

4. *How sensitive is the decision to changes in the estimates?* That is, if, because of uncertainty, an estimate of risk or benefit were wrong by a factor of 2, or 10, or 100, would the decision maker's choice be different?

5. *What other risk and risk control assessments* have been made, and why are they different from those now being offered?

Information on Management

"Risk management" is a term used to describe processes surrounding choices about risky alternatives. In common usage, assessments of the risks and benefits of various options are seen as technical activities that yield information for decision makers, whose decisions are called risk management decisions (National Research Council, 1983a). [If one accepts the distinction between risk assessment and risk management (see the list of terms in Appendix E), communication about risks that involves nonexperts would generally be part of risk management.] In addition to information about risks and benefits, decision makers need answers to managerial questions such as these:

1. *Who is responsible for the decision?* Who is responsible for preventing, mitigating, or compensating for damage? Who is responsible for generating and evaluating data? Who has oversight?

2. *What issues have legal importance?* Do the applicable laws take benefits into consideration? Do they allow consideration of the risks of alternatives? Do they require the analysis of economic and social impacts of the activity in question or its alternatives?

3. *What constrains the decision?* What technical, physical, biological, or financial limits constrain some possible choices? What are the limits of authority of the person or organization making the decision? Are there time limits imposed on the decision process? What difference could public opinion or political intervention make?

4. *What resources are available for implementing the decision?* What personnel and financial resources are available to the decision maker? To others involved in debating the decision?

Other Relevant Knowledge

In addition to items on the above lists, other considerations are also important. Technological choices involve risks and benefits not only to the life, health, and safety of individual humans but also to nonhuman organisms, ecological balances, the structures of human communities, political and religious values, and other things that concern decision makers but that are not easily evaluated by the quantitative approaches implied by the above lists. The assessment of such risks and benefits is not standard practice in the field of risk assessment. Such factors are commonly discussed, however, in activities and documents described as "impact assessments" or "technology assessments." These broadly conceived activities and documents often address a wide range of the questions just outlined.

Summary

In sum, *a well-informed choice about activities that present hazards and risks requires a wide range of knowledge.* It depends on understanding of the physical, chemical, and biological mechanisms by which hazardous substances and activities cause harm; on knowledge about exposures to hazards or, where knowledge is incomplete, on analysis and modeling of exposures; on statistical expertise; on knowledge of the economic, social, esthetic, ecological, and other costs and benefits of various options; on understanding of the social values reflected in differential reactions to the qualities of risks; on knowledge of the constraints on and responsibilities of risk managers; and on the ability to integrate these disparate kinds of knowledge, data, and analysis. Needless to say, *it is often impossible in practice to gather all this knowledge.* Nevertheless, the more complete the knowledge and the more quantitative answers are found, the better informed the ultimate decision will be.

GAPS AND UNCERTAINTIES IN KNOWLEDGE

The above summary of needed knowledge clearly suggests that decisions about risky activities and hazardous substances are frequently made with incomplete information. In this section we elaborate on some of the points just raised. We focus on risks, even though there are major gaps and uncertainties in knowledge about benefits as well, and we list several important ways that information about

"*Then we've agreed that all the evidence isn't in, and that even if all the evidence were in, it still wouldn't be definitive.*"

FIGURE 2.1   SOURCE: Drawing by Richter; ©1987 The New Yorker Magazine, Inc.

the nature and magnitude of risk is often incomplete and uncertain (see Figure 2.1).

Identification of Hazards

It is sometimes difficult even to determine whether a hazard exists. For activities or substances whose hazards are delayed in onset (such as possible causes of cancer or birth defects) and for substances to which people are exposed in very small quantities, it is difficult to connect effects to causes. Analysts often use experiments with animals or bacteria to determine whether such activities or substances are hazardous under controlled conditions, but not all potential hazards are studied, even in the laboratory. A National Research Council panel reviewed the testing that had been done on a random sample of 675 substances (National Research Council, 1984). Within this group, 75 percent of the drugs and inert chemicals in drug formulations had had some testing for acute toxicity and 62 percent had had some testing for subchronic effects. For pesticides and ingredients in pesticide formulations, these values were 59 percent and 51 percent, respectively. Testing for chronic, mutagenic, or reproductive and developmental effects was less frequently done than testing for acute and subchronic effects, and testing of all kinds was less frequently done for substances on the Toxic Substance Control Act's list of chemicals in commerce. The panel concluded that toxicity studies had not yet been done on the majority of the chemicals—amounting to tens of thousands—now in industrial use in the United States.

Even when studies have been done with lower organisms, it is uncertain whether there is a human hazard. Substances that cause cancer, mutations, or birth defects in some species of animals often have no demonstrable effect on other species, and the reasons for these differences are not yet understood. For instance, a review by the Food and Drug Administration indicated that of 38 compounds demonstrated or suspected to cause birth defects in humans, all except one tested positive in at least one animal species and more than 80 percent were positive in more than one species. Eighty-five percent of the 38 compounds caused birth defects in mice, 80 percent in rats, 60 percent in rabbits, 45 percent in hamsters, and 30 percent in primates (National Research Council, 1986b). Thus some substances that do not cause cancer or birth defects in test species appear to have these harmful effects on humans. And the reverse may also be true. Scientists may agree that positive results in an animal test on a particular substance are strong evidence of a human hazard, but there is always some uncertainty about that judgment.

Estimation of Exposure

Data are frequently inadequate on exposures to hazards. Many hazardous substances are diffused in the air or in surface or underground waterways and in the process undergo physical or chemical changes that transform them into other substances that may be less hazardous—or that may be more so, although more dilute. Many hazardous substances are transformed by biological processes before they reach humans. And even in the human body, metabolic processes can alter hazardous chemicals before they reach the organs to which they present hazards, sometimes making them less toxic, but sometimes making them more so (National Research Council,

1986b). Thus the hazardous substances released into the environment at the source may be very different in quantity and even in kind from those to which people are ultimately exposed. The measurement of exposure is therefore most accurate at the dispersed sites where people live and work. As a result, it can be very expensive to collect accurate exposure data. The problems and the expense multiply when researchers try to address questions about unequal distributions of exposure and about possibly sensitive populations. Many more measurements must be made to compare the exposures of a variety of populations. For these reasons exposures are usually estimated from data on releases of hazardous substances. Inferring exposures requires numerous assumptions about the transport, dispersion, and transformations of substances, many of which are based on incomplete theory and limited evidence (National Research Council, 1988a). The use of estimates rather than measurements of exposure adds a layer of uncertainty to risk estimates.

Further uncertainty is introduced by the fact that many hazards produce their effects by exposure over time. It is known that exposure to radiation and some hazardous substances in a given amount will have different effects depending on whether it occurs at once, is spread over several smaller exposures, or is continuous at a low rate over a long period of time (National Research Council, 1988b). It is not known, however, how much difference this time dimension makes for particular hazards or which rate of exposure carries the greatest risk (National Research Council, 1984:60).

Estimation of the Probability of Harm

Knowledge about the probability of harm from a given hazard is also frequently inadequate or uncertain. The best way to estimate the probability of harm is to examine the accumulated experience of people exposed to the hazard. Only rarely, however, as with automobile travel and other familiar hazards whose effects are easy to observe, is there sufficient human experience to calculate accurate probabilities from observational data. Past experience does not exist for many controversial hazards because they involve new technologies. For many others, including carcinogens and most air pollutants, past experience is hard to interpret because it is difficult to tell which illnesses or deaths are attributable to the hazard rather than to other causes. For yet other hazards the meaning of past experience is in dispute because the greatest concern is about very low probability but potentially disastrous events, such as a nuclear reactor core melt-down or the escape of a virulent organism from a laboratory. The fact that a disaster has not happened may mean that there is no potential for harm, that the potential is high but luck has been good, or that the probability of harm is very low. But when considering major disasters, even a very low probability can mean the risk to the population, defined as the probability multiplied by the magnitude of the consequence, is large.

When knowledge from experience is unavailable or unreliable, analysts develop methods of estimating the risk. To assess the risk from carcinogens, they commonly use data from laboratory experiments on nonhuman organisms. Adding assumptions about how humans differ from the experimental organisms and about how to extrapolate from the 2-year exposures to high doses usually given to laboratory rodents to the long-term low doses characteristic of natural human exposures, they estimate the human risk. An extensive literature debates the merits of different methods of making these extrapolations across species, dosages (National Research Council, 1980), and exposure times (Kaufman, 1988). Risk analysts also use epidemiological studies that correlate evidence of exposure and evidence of harm, but interpretations of these studies are often controversial because they are open to alternative explanations. For instance, illnesses in exposed groups may be due to some other hazard to which they were also exposed or to some synergistic interaction of hazards. Only very infrequently do analysts have access to data on humans whose exposures to the relevant hazards are well known.

A different sort of uncertainty arises in assessing the risk of disasters that result from the breakdown of complex technological systems, particularly types of catastrophic accidents that have not previously occurred. Risk analysts sometimes address this problem with "fault-tree" analysis, a technique that uses experience to estimate the probabilities of various events that might contribute to a disaster and then combines the probabilities to estimate the likelihood that enough contributing factors will occur at once to trigger the disaster. The analysts then use available data and models to estimate potential exposures and their consequences. Needless to say, these methods of estimation are full of untested assumptions and uncertainties. In particular, an extensive literature debates the errors of omission and commission in fault-tree analyses of the probability of technological disasters, such as in the nuclear power industry (Campbell and Ott, 1979; Fischhoff et al., 1981a; McCormick, 1981).

FIGURE 2.2   SOURCE: Drawing by Richter; ©1988 The New Yorker Magazine, Inc.

The uncertainties in these methods are legion, so several different and even conflicting conclusions can often be defended by competent scientists. It is difficult and sometimes proves impossible to reach a consensual judgment about what the probabilities are, let alone what to do about the attendant risks (see Figure 2.2).

Identification of Synergistic Effects

Additional uncertainty in risk estimates exists because exposure to one hazard can affect a person's sensitivity to another. For instance, asbestos is estimated to be about 10 times as dangerous to smokers as to nonsmokers (Breslow et al., 1986). This may occur because chemical reactions between the substances yield products of different toxicity or because one substance increases the availability to the body of another one that would not have been toxic by itself (National Research Council, 1988a). In such ways, exposure to one substance can potentiate the adverse effects of another or, less commonly, decrease another substance's toxic effect. There is very little knowledge, however, about how frequent or how strong such synergistic or blocking effects are or about which combinations of substances and activities are likely to exhibit the effects. The knowledge that such effects exist, however, gives reason to consider almost all estimates of health risk based on studies of single hazardous substances as somewhat uncertain, even when they are based on the most careful analysis possible.

Summary

In sum, *any scientific risk estimate is likely to be based on incomplete knowledge combined with assumptions, each of which is a source of uncertainty that limits the accuracy that should be ascribed to the estimate.* Does the existence of multiple sources of uncertainty mean that the final estimate is that much more uncertain, or can the different uncertainties be expected to cancel each other out? The problem of how best to interpret multiple uncertainties is one more source of uncertainty and disagreement about risk estimates.

SCIENTIFIC JUDGMENT AND ERRORS IN JUDGMENT

What do analysts do when confronted with knowledge so full of uncertainties? Scientists' training, which teaches them to accurately represent certain types of uncertainties, comes into conflict with the pressure to give succinct, unambiguous answers that can inform the social and personal decisions nonexperts must make about risks. If the experts remain silent or equivocal, choices will be made without taking into account what they know. Once they begin to convey what they know, however, experts must inevitably make judgments about the meaning of available information and about the degree to which uncertainty makes it less reliable. But because experts rely on ordinary cognitive processes to make sense of the wealth of data they have available, their judgments about the meaning and conclusiveness of available information can suffer from some of the same frailties that affect human cognition in general.

Inappropriate Reliance on Limited Data

Even statistically sophisticated individuals often have poor intuitions about how many observations are necessary to support a reliable conclusion about a research hypothesis (Tversky and Kahneman, 1971). In particular, they tend to draw conclusions from small samples that are only justified with much larger samples. Thus they may be prone to conclude that a phenomenon such as a toxic effect does not exist when in fact the data are so sparse that the only appropriate conclusion is that the search for the phenomenon is in its early stages. They may also err in the opposite direction, sounding an alarm on the basis of extremely limited preliminary data. The tendency for scientists to draw conclusions from "low-power" research has been documented in fields from psychology (Cohen, 1962) to toxicology (Page, 1981). Low-power research uses measurements and methods that are unlikely to reveal small effects without very large numbers of measurements. Where the tendency to premature conclusion operates, expert judgment will err by underreporting or overreporting effects, both hazardous and beneficial.

Tendency to Impose Order on Random Events

People who are seeking explanations for events, including experts working in their areas of expertise, have a tendency to see meaning even when the events are random (Kahneman and Tversky, 1972). For instance, stock market analysts develop elaborate theories of market fluctuations, but their predictions rarely do better than the market average (Dreman, 1979), and clinical psychologists see patterns they expect to find even in randomly generated test data (O'Leary et al., 1974). In interpreting statistics relating the incidence of cancer to occupational exposures to particular chemicals, there is a temptation to interpret a correlation between exposure to a particular chemical and the incidence of a particular cancer as evidence of an effect. But some such evidence is to be expected even in random data, if large numbers of chemicals and cancers are examined. Similarly, occasional "cancer clusters" are likely to be present in large epidemiological studies even by chance. Replication on a new sample is the best way to check the reliability of such relationships, but new samples are often hard to find. Sometimes, conclusions are reported and publicized as definite before they have been adequately checked.

Such instances, including the interpretation of "unusual" cases, are at heart issues of the proper conduct of scientific analysis. Although recent attention on scientific misconduct may attach greater significance to unusual cases than is actually warranted, it is nonetheless important to recognize the natural human tendency to find order even when the evidence is tenuous and to recognize that when analysts are strongly motivated to find particular results they may overinterpret the evidence.

Tendency to Fit Ambiguous Evidence into Predispositions

When faced with ambiguous or uncertain information, people have a tendency to interpret it as confirming their preexisting beliefs; with new data they tend to accept information that confirms their beliefs but to question new information that conflicts with them (Ross and Anderson, 1982). Because of the high degree of ambiguity in the data underlying risk assessments, this cognitive bias may act to perpetuate erroneous early impressions about risks even as new evidence makes them less tenable.

Tendency to Systematically Omit Components of Risk

In analyses of complex technological systems, certain features are commonly omitted, possibly because they are absent from operating theories of how the technological systems work. In particular, analysts are prone to overlook the ways human errors or deliberate human interventions can affect technological systems; the ways different parts of the system interact; the ways human vigilance may flag when automatic safety measures are introduced; and the possibility of "common-mode failures," problems that simultaneously affect parts of the technological system that had been assumed to be independent [for elaboration and citations of the evidence, see Fischhoff et al. (1981a)]. Typically, people who were not involved in performing the analyses are unlikely to notice such omissions—in fact, in a complex technical analysis, observers are likely to overlook even major omissions in the analysis. Although most of these oversights tend to lead to underestimates of overall risk, this need not always be the case.

Overconfidence in the Reliability of Analyses

Weather forecasters are remarkably accurate in judging their own forecasts. When they predict a 70 percent chance of rain, there is measurable precipitation just about 70 percent of the time. They seem to be so successful because of the following characteristics of their situation: (1) they make numerous forecasts of the same kind, (2) extensive statistical data are available on the average probability of the events they are estimating, (3) they receive computer-generated predictions for specific periods prior to making their forecasts, (4) a readily verifiable criterion event allows for quick and unambiguous knowledge of results, and (5) their profession admits

its imprecision and the need for training (Fischhoff, 1982; Murphy and Brown, 1983; Murphy and Winkler, 1984). Most of these conditions do not hold for professional risk assessors, however, and the predictable result is overconfidence among experts. For instance, civil engineers do not normally assess the likelihood that a completed dam will fail, even though about 1 in 300 does so when first filled with water (U.S. Committee on Government Operations, 1978).[2]

Summary

These normal cognitive tendencies can lead expert risk analysts to convey incorrect impressions of the nature and reliability of scientific knowledge. Some of the tendencies predispose to premature judgment that a risk is low or high. Several of them bias scientific judgment in the direction of overconfidence about the certainty of whatever currently seems to be known. Although the net effect of these cognitive tendencies has not been determined, their existence justifies a certain amount of skepticism on the part of decision makers, including individuals, about definitive claims made by risk analysts.

INFLUENCES OF HUMAN VALUES ON KNOWLEDGE ABOUT RISK

Although it is useful conceptually to separate risk assessment and risk control assessment from value judgment, there are many respects in which it is not possible to accomplish the separation in practice. Judgments made by scientists on which types of hazardous consequences to study and by analysts on which ones to measure are based in part on technical information—what knowledge already exists, what additional knowledge would be relevant to a decision at hand, what the relative costs are of collecting different kinds of data, and what kinds of information would be most useful for estimating particular risks. But they are also based on value judgments about which types of hazard are most serious and therefore most worthy of being reduced. This section discusses two of the ways that human values enter understanding of risks: through the choice of numbers to summarize knowledge about the magnitude of risks and through the weighting of different attributes of hazards.

Choices of Numerical Measures for Risk

The need to quantify risks as an aid to decision making creates special difficulties because the choice of which numerical measure to use depends on values and not only on science. This fact is evident even in a simple problem of risk measurement—the choice of a number to summarize information on fatalities. Different risk analysts have used different summary statistics to represent the risk of death from an activity or technology.[3] Among the measures used are the annual number of fatalities, deaths per person exposed or per unit of time, reduction of life expectancy, and working days lost as a result of reduced life expectancy. The choice of one measure or another can make a technology look either more or less risky. For instance, in the period from 1950 to 1970, coal mines became much less risky in terms of deaths from accidents per ton of coal, but they became marginally riskier in terms of deaths from accidents per employee (Crouch and Wilson, 1982). This is because with increasing mechanization fewer workers were required to produce the same amount of coal. So although there were fewer deaths per year in the industry, the risk to an individual miner actually increased during this period. Which measure is more appropriate for decisions depends on one's point of view. As some observers have argued, "From a national point of view, given that a certain amount of coal has to be obtained, deaths per million tons of coal is the more appropriate measure of risk, whereas from a labor leader's point of view, deaths per thousand persons employed may be more relevant" (Crouch and Wilson, 1982:13).

Each way of summarizing deaths embodies its own set of values. For example, "reduction in life expectancy" treats deaths of young people as more important than deaths of older people, who have less life expectancy to lose. Simply counting fatalities treats deaths of the old and young as equivalent; it also treats as equivalent deaths that come immediately after mishaps and deaths that follow painful and debilitating disease or long periods during which many who will not suffer disease live in daily fear of that outcome. Using "number of deaths" as the summary indicator of risk implies that it is equally important to prevent deaths of people who engage in an activity by choice and deaths of those who bear its effects unwillingly. It also implies that it is equally important to protect people who have been benefiting from a risky activity or technology and those who get no benefit from it. One can easily imagine a range of arguments to justify different kinds of unequal weightings for different kinds of deaths, but to arrive at any selection requires a value judgment concerning which deaths one considers most undesirable. To treat the deaths as equal also involves a value judgment.

There are additional value choices involved in calculations based on fatalities. A particularly controversial choice concerns whether to "discount" lives, that is, whether to give deaths far into the future less weight than present deaths. This approach to valuation is sometimes advocated on the ground that people typically prefer a given amount of any particular good in the present to the same value in the future—if they invested the cost of the good, they could expect to have increased purchasing power and thus to be able to purchase more of it in the future than in the present. Although one cannot "invest" human life in the same way, society can invest the resources used to save or prolong lives. From an individual's point of view, one arguably loses less by dying at an old age than when younger, so people may be less willing to work to avoid probable deaths the farther they are in the future.

Discounting is controversial partly because it is used to put a monetary value on human life. Some measure, whether based on probable future earnings or consumption or on willingness to pay to reduce the probability of fatality, is selected to put a price on what for many has intrinsic moral or even religious value—and each of these measures embodies controversial assumptions about what is worthwhile about life. In addition, choosing a positive discount rate—one that treats future lives as worth less than present lives— suggests that society cares less about its children's generation than its own, a controversial assumption to say the least. But deciding not to discount lives also involves a judgment about the future, and so it is also a value-laden choice (Zeckhauser and Shephard, 1981).

Values also enter into scientists' choices about how to characterize the uncertainty in their information. It is traditional among civil engineers, public health professionals, and others to take account of uncertainty by being "conservative" in stating risk estimates. This means that they leave a margin for error that will protect the public if the actual risk turns out to be greater than the best currently available estimate. But it has sometimes been argued that risk analysts should instead present their best available estimate to decision makers, along with an explicit characterization of its uncertainty, and allow the decision makers to decide explicitly how much margin of safety to allow. The dispute is highly controversial because many

"HEY, I THOUGHT WE WERE WORKING WITH THE SAME DATA..."

FIGURE 2.3   SOURCE: *National Wildlife Magazine*, August–September, 1984.
Copyright © 1984 Mark Taylor. Reprinted with permission of Mark Taylor.

believe that in practice the latter approach will provide a narrower
margin of safety. The central point here is that either way of representing uncertainty embodies a value choice about the best way to
protect public health and safety.

These few examples show how human values can enter into even
apparently technical decisions in risk analysis, such as about the
choice of a number to summarize a body of data. It is easy therefore
to see how choices that are justified by appeal to data from a risk
analysis can sometimes be questioned by appealing to the very same
data (see Figure 2.3).

Values and the Attributes of Hazards

We have noted that decision makers do not choose among risks
but among alternatives, each with many attributes, only some of
which concern risk. Similarly, each hazard—and, for that matter, each benefit—that a decision alternative presents has many
attributes. These attributes are important to nonexperts for the
purpose of making decisions.

Qualitative aspects of hazards are relevant to decisions in various ways. In different decision contexts it may be necessary to consider comparisons and trade-offs such as the following: Is a risk of cancer worse than a risk of heart disease? Is an accidental death of a person at age 30 more to be avoided than a death by emphysema at age 70? Is an industrial hazard more acceptable if it is borne by workers partly compensated by their pay than if it is borne by nonworking neighbors of the industrial plant? Are the deaths of 50 passengers in separate automobile accidents equivalent to the deaths of 50 passengers in one airplane crash? Is a hazard that faces the unborn worse than a similar hazard that we face ourselves? Is a large hazard with a low probability equally undesirable as a small hazard with a high probability when the estimated risks are equal? The difficult questions multiply when hazards other than to human health and safety are considered. Technological choices sometimes involve weighing the value of a river vista, a small-town style of living, a holy place, or the survival of an endangered species, in addition to dangers to human health, against probable economic benefits. Such choices are ultimately matters of values and interests that cannot be resolved merely by determining what the risks and benefits are.

A growing body of knowledge on what is usually called "risk perception" helps illuminate the values involved in the evaluation of different qualities of hazards.[4] In studies of risk perception individuals are given the names of technologies, activities, or substances and asked to consider the risks each one presents and to rate them, in comparison with either a standard reference or the other items on the list. The responses are then analyzed, taking into account attributes of the hazards and benefits each technology, activity, or substance presents (Table 2.1 lists several such attributes). Analysis consistently shows that people's ratings are a function not only of average annual fatalities according to the best available estimates, but also of the attributes of the hazards and benefits associated with a technology, activity, or substance (Fischhoff et al., 1978; Gould et al., 1988; Otway and von Winterfeldt, 1982; Slovic et al., 1979, 1980). In particular, the studies show that certain attributes of hazards, such as the potential to harm large numbers of people at once, personal uncontrollability, dreaded effects, and perceived involuntariness of exposure, among others (see Table 2.1), make those hazards more serious to the public than hazards that lack those attributes. Also, choices that provide different types of benefit, such as money, security, and pleasure, are valued differently from each other (Gould et al., 1988). The fact that hazards differ dramatically in their qualitative aspects helps explain why certain technologies or activities, such as nuclear power, evoke much more serious public opposition than others, such as motorcycle riding, that cause many more fatalities.

An important implication of such findings is that those quantitative risk analyses that convert all types of human health hazard to a single metric carry an implicit value-based assumption that all deaths or shortenings of life are equivalent in terms of the importance of avoiding them. The risk perception research shows not only that the equating of risks with different attributes is value laden, but also that the values adopted by this practice differ from those held by most people. For most people, deaths and injuries are not equal— some kinds or circumstances of harm are more to be avoided than others. One need not conclude that quantitative risk analysis should weight the risks to conform to majority values. But the research does suggest that it is presumptuous for technical experts to act as if they know, without careful thought and analysis, the proper weights to use to equate one type of hazard with another. *When lay and expert values differ, reducing different kinds of hazard to a common metric (such as number of fatalities per year) and presenting comparisons only on that metric have great potential to produce misunderstanding and conflict and to engender mistrust of expertise.*

IMPLICATIONS FOR RISK COMMUNICATION

We have shown in this chapter that different experts are likely to see technological choices in different, sometimes contradictory, ways even when the information is not at issue. Incomplete and uncertain knowledge leaves considerable room for scientific disagreement. Judgments about the same evidence can vary, and both judgments and the underlying analyses can be influenced by the values held by researchers. Since scientists and the people who convert scientific information into risk messages do not all share common values, it is reasonable to expect risk messages to conflict with each other. Even in the best of circumstances for communication, conflicting risk messages would create confusion in the minds of nonexperts who must rely on them to inform their choices. But as the next chapter shows, the circumstances are not the best. The social conflict that surrounds modern technological choices is characterized by anxiety and mistrust and by clashes of vested interests and values, conditions that create formidable tasks for those who would improve decision making through risk communication.

NOTES

1. One technical definition of risk is that risk is the product of a measure of the size of the hazard and its probability of occurrence. Regardless of how numerical estimates are made, the essence of the distinction between hazard and risk is that "risk" takes probability explicitly into account.

2. This discussion is drawn from Fischhoff et al. (1981a). More extensive discussions of expert overconfidence with additional examples can be found there and in Lichtenstein et al. (1982).

3. This discussion is drawn from Fischhoff et al. (1984:125–126), where further citations can be found.

4. The term "risk perception" is put in quotation marks because, as the discussion shows, this body of research is more accurately described as the study of human values regarding attributes of hazards (and benefits).

3
Conflict About Hazards and Risks

Conflict within our society about technological choices, focusing on hazards and risks, is an essential part of the environment in which those choices are debated and made (e.g., Dickson, 1984; Lawless, 1977; Mazur, 1981; Nelkin, 1979a).[1] That is, conflict is an essential part of the environment of risk communication. This chapter discusses the reasons communication about hazards and risks in the U.S. political system has become so contentious over the last two decades. It identifies the major sources of this increasing conflict and briefly explores the nature of that conflict. Risk communication is profoundly affected by the conflictual atmosphere in which it occurs.

IS RISK INCREASING OR DECREASING?

For many observers the central dispute about technology and risk concerns whether risk is increasing or decreasing (e.g., National Research Council, 1982). In some accounts people are concerned about the risks of technology because there is an increasing threat of technological disaster; in other accounts, public concern flies in the face of a demonstrable decrease in net risk to human health and survival. Although we do not believe this debate to be productive for risk communication, a brief and simplified account of it will serve to introduce the discussion that follows, concerning the sources of increasing conflict about technological choices.

TABLE 3.1 Life Expectancies in the United States, 1900-1984

	White Male	Black Male[a]	White Female	Black Female[a]
Life Expectancy at Birth				
1900-1902	48.2	32.5	51.1	35.0
1949-1951	66.3	58.9	72.0	62.7
1984	71.8	65.6	78.7	73.7
Remaining Life Expectancy at Age 25				
1900-1902	38.5	32.2	40.1	33.9
1949-1951	44.9	39.5	49.8	42.4
1984	48.7	43.1	55.0	50.7

[a]Life expectancy figures for 1949-1951 are for nonwhites.

SOURCE: Metropolitan Life Insurance Company Statistical Bulletin, 1987.

It Is the Safest of Times

Proponents of the view that this is the safest of times[2] point out that the best overall measure of health and safety risk is average life expectancy. They note that during this century there have been dramatic increases in life expectancy even as the society has increased its use of the chemicals and other hazardous substances that are the subject of intense debate about risk. The increases have been marked for women and men and for blacks and whites (see Table 3.1). While much of the increased longevity is due to declining infant mortality and is probably unrelated to environmental and occupational health hazards, improvements in life expectancy of young adults have also been striking. Thus medical science, improved nutrition, water purification, and other advances have combined to give each person a good chance at living a full life span. The data offer no indication that epidemics of chemical-induced cancer or other technologically borne scourges are increasing the risk of fatality.

Proponents of the view that risk is decreasing point out that many of the hazardous substances now in the environment decrease overall risk by replacing more dangerous substances. For instance, chlorinated hydrocarbon solvents, which cause cancer in animals and possibly humans as well, have replaced flammable ones, which caused death by fire. Many other hazardous substances decrease risk by reducing more serious preexisting hazards. Pesticides and herbicides may cause cancer, but, in some parts of the world at least, they have helped prevent famine. Water chlorination increases exposure to carcinogens but decreases exposure to typhoid-causing bacteria and other infectious agents.

Proponents of the view that technology improves safety conclude that many people are becoming more and more concerned about smaller and smaller risks. They see the gains from past technological change as outweighing the new risks by a large margin, and they see no reason the trend will not continue.

It Is the Riskiest of Times

Proponents of the view that this is the riskiest of times see modern technology as generating new threats to society and the earth's life-support systems and as doing so at an accelerating pace. They argue that because of the technological advances that have increased life spans, population growth threatens more devastating famines than the world has ever seen. They also note that the long-term biological and ecological effects of rapid increases in the use of chemicals are still unknown. To illustrate the reason for concern, they note that serious hazards continue to be discovered—a recent example is the hazard to the earth's ozone layer from manufactured chlorofluorocarbons. They point out that the synergistic effects of technological hazards remain almost entirely unstudied even though people are rarely exposed to one hazard in isolation from others. They point to a range of global environmental threats whose ultimate implications for humanity are unknown but potentially catastrophic: the rapid rate of extinction of species and the destruction of their habitats; deforestation and decreases in biological diversity in the tropics; the possibility of major climatic change due to human activity; and, of course, the possibility of nuclear holocaust.

Proponents of the view that technology is increasing risks do not see advances in life expectancy as a convincing counterargument. They point out that many of the new risks are unlikely to be reflected in current life expectancy data because they are so far only evident in indicators of ecosystems and the geosphere. They note that the new low-probability catastrophic risks that they consider important cannot appear in life expectancy tables because the catastrophes have not yet occurred. And they suggest that progress in raising life expectancy, which has slowed since 1950, might have been greater if it had not been for the new risks. Thus those who see risk as increasing call for tighter control over technology, introduction of more environmentally benign technology, and abandonment of some technologies considered particularly risky.

Understanding the Conflict

Although each of these views has some valid and convincing evidence on its side, the dispute cannot be resolved by available evidence. In fact, it may not ultimately be about evidence. At a deeper level it is about what kinds of risks people want most to avoid, what kinds of lives they want to lead, what they believe the future will bring, and what the proper relationship is between humanity and nature. Reviewing the evidence will not resolve the dispute— in fact, debates over technology framed in this way seem only to increase anger and frustration. But understanding the conflict may be a necessary first step toward improving dialogue, that is, toward making better risk communication possible.

To understand the conflict, it helps to begin by asking what has changed in the relation of technology and society and what has not. As we noted in Chapter 2, the existence of technological hazards is nothing new. Whether such hazards present an increased net risk is, of course, a matter of dispute. There is little doubt, however, that the extent and intensity of conflict about technological hazards have increased substantially over the past 30 years. This can be seen in the pressures that culminated in a flurry of environmental legislation in the late 1960s and the 1970s, in evidence of increasing public opposition to nuclear power since the early 1970s (Ahearne, 1987; Freudenburg and Rosa, 1984; Hively, 1988), and in the continuing strong public support for environmental regulation during the Reagan years in the face of the administration's commitment to deregulation (Dunlap, 1987).[3] The following sections elaborate on the major factors contributing to intense conflict over technology and on the nature of that conflict.

CHANGES IN THE NATURE OF HAZARDS AND IN KNOWLEDGE ABOUT THEM

The hazards recognized in modern living have changed in kind, regardless of whether any particular type of risk has increased or decreased. In addition, new knowledge about hazards and risks has led people to think about them in new ways. The important changes described below give reason for a continuing high level of public concern (Dunlap, 1987; Mitchell, 1980).

Increased Understanding of Human Influence on Hazards

Advances in science and technology have made clear that humanity has much more to do with its own health and longevity than was once believed. Many illnesses and deaths that were once seen as inevitable, random, or divinely caused are now known to have human origins. Modern science can detect anthropogenic toxic substances at increasingly low concentrations and can trace their biological effects with animal experiments and epidemiological studies. Modern techniques of detection and analysis can connect events over great distances and through complex pathways, revealing the human causes of hazards.

People are also increasingly aware that human action can avoid or reduce risks. Individuals have learned that they can increase their life expectancies by wearing seat belts, avoiding tobacco use, and controlling their diets. Governments and firms can reduce human health risks with pollution controls and improved safety measures in industrial processes and consumer products. And, of course, medical science continues to develop ways to prolong life. It is an irony of progress that each success in prolonging and enhancing human life brings increasing awareness that human action—or inaction—can also be responsible for death.

Awareness of the human influence over life and death makes technological choices into moral issues. In most modern societies harm to a person readily becomes a moral issue if a responsible party can be identified. Thus people feel morally obligated to donate blood or bone marrow when they are made to understand that their particular type is needed to prolong life (Schwartz, 1977). Similarly, people who believe industrial firms are responsible for some cancers tend to see them as morally obligated to ameliorate the harm (Stern et al., 1986). From such moral feelings comes the widespread sentiment for using extraordinary, risky, and expensive measures to prolong lives when nothing else is likely to work. By the same reasoning, reports that the burning of coal in Ohio is killing fish in New York and may be threatening human health can lead people to see the pollution of air as immoral.

In the U.S. and other legal systems, awareness of human influence calls into action fundamental norms about responsibility, rights, and due process. When people who are perceived to be innocent are put in jeopardy, discussions about intent, justice, blame, and punishment are almost inevitable. What is at issue is no longer only whether an activity makes people better or worse off but whether the changes are fair and whether the responsible agent has the right to affect other people's well-being.

Worsening Worst Cases

Modern technology, by making it possible for humans to alter natural processes at the level of the geosphere, has made possible disasters that could not even be fantasized a few generations ago. Already, deforestation is disrupting huge ecosystems, and there is evidence that it, combined with the burning of unprecedented quantities of fossil fuel, is altering the earth's temperature and threatening to raise the level of the oceans and disrupt the patterns of temperature and precipitation on which world agriculture depends. Although deforestation leading to climatic disruption is not new—it is responsible for the present aridity in much of the Middle East and China—human alteration of climate has never before been possible on a global scale. There is dispute over the probability of a climatic catastrophe, but little dispute that global climatic changes of historic proportions are now possible as a result of human activity (Jaeger, 1988). Similarly, the threat to the earth's ozone layer suggests the possibility of human-generated environmental damage on an unprecedented scale. And, of course, the possibility of devastation of whole nations by nuclear weapons is unprecedented.

Most of the unprecedented catastrophes scientists have described have a very low probability of occurrence, but because the outcomes are so undesirable the risks are worth considering carefully. However, the low probability makes them hard to analyze. An example is major disasters from nuclear power plant operation. The industry is too young for the probability to be estimated accurately from experience; yet indirect methods of estimation are highly uncertain. Thus people are left with huge disasters to contemplate but no reliable guidance about how seriously to take them.

With worsening worst cases, it makes sense to pay attention to smaller and smaller probabilities and to smaller differences between probability estimates. But most people have difficulty understanding very low probabilities (see, e.g., Fischhoff et al., 1981b). They tend to think in the categories of language (such as "never," "rarely," "occasionally," "often," and so forth) rather than along the continuous dimensions of mathematics (cf. Starr and Whipple, 1980). For very low probability events, nonexperts tend to use two categories, "possible" and "effectively impossible." Thus the changes that have made nightmares into possibilities may drastically alter many people's thinking by making a qualitative change—by making them aware of a hazard where they had perceived none. People may pay more attention to the size of the consequences and ignore both the magnitude and the uncertainty of very low probability estimates. The result would be a much-increased concern about catastrophic risks and a corresponding increase in opposition to technologies that pose them.

Unintended Side Effects

Technological activity has probably always had effects on people who were not directly involved in it, but knowledge of the extent of such effects has increased dramatically in this century. Technological changes are accelerating, as are the materials and energy transformations that can disturb preexisting physical and biological systems and affect human well-being. Although people have always been exposed to the side effects of other people's activity, they are now aware of being exposed to much more and at greater distances. There is increasing evidence that technological activities can now affect people around the earth by altering air quality, exposing them to ultraviolet radiation, or changing climate.

When side effects spread more widely and when that change is recognized, collective action often follows. The risk bearers tend to take up common interest against the risk givers. And when the effects extend across the boundaries of communities and then of nations, the conflicts of interest often enter formal political and diplomatic arenas or, if those are not available, find informal ways of gaining wide attention. Thus increasing technological conflict is due in part to the widening range of technology's effects and the greater social awareness of the change.

Changing Portfolio of Hazards

The hazards society confronts today are different from those of the past. As noted in Chapter 2, the principal threats to health, especially among the more educated and politically active segments of the public, are now from chronic diseases rather than acute illnesses and from illnesses now known to have long latency periods. Sometimes decades pass between exposure and effect; sometimes the effect manifests itself only in later generations. Whereas infectious diseases can be convincingly linked to microorganisms in the body, cancer and many other chronic diseases cannot, in general, be conclusively linked to causative agents.[4] People are often unsure what caused such illnesses. Moreover, if they are exposed to a hazard, they cannot know whether they will become ill. People spend more of their lives under a cloud: whenever they are exposed to a "probable carcinogen" or other hazard with delayed potential effects, they may worry about whether it will eventually harm them. If they become ill, they can consider a range of hypotheses about human actions that might have been to blame: past occupational exposure, dietary practice, air pollution, and so forth. Some people agonize over whether they are guilty of causing their own illness; others conclude that they are innocent victims of greed or negligence. The former conclusion produces anxiety; the latter, whether correct or not in any particular instance, motivates lawsuits and other forms of social conflict.

Hazards have also changed in that there is more knowledge—and more widespread awareness—of hazards to which people are exposed but over which they have no control as individuals. Individuals on their own are helpless to reduce the risks of nuclear war, depletion of the ozone layer, and global climatic change. Media accounts make people acutely aware of other hazards that strike more or less at random, such as airplane hijackings and releases of toxic substances such as at Bhopal or radioactivity such as at Chernobyl. People have learned that some industrial chemicals are toxic but that for many chemicals now widely used in commerce in the United States little is known about whether they threaten human health (National Research Council, 1984). The anxiety that comes from awareness of apparently uncontrollable risks derives in large part from a sense of uncertainty. People may get the sense that past experience—including longevity tables—may not provide a reliable estimate of the risks they face.

For highly uncertain risks it is difficult to refute extreme estimates of their magnitude. Concerns may persist precisely because of the uncertainty. An example is the concern that AIDS may be transmitted by mosquitoes. While technical experts agree that mosquito transmission is too improbable to worry about, a skeptic can maintain that it has not been proven impossible. Additionally, highly uncertain risks generate special conflicts about their management, with decision makers disagreeing widely about how large a margin of safety should be allowed to protect against the occurrence of disastrous consequences that they agree are unlikely.

CHANGES IN U.S. SOCIETY

Technological decisions have become more controversial in part because U.S. society has changed in several ways in the era since World War II.

Increasing Affluence

For most of those who participate actively in American politics, economic security has allowed certain basic human concerns to recede from awareness and to be replaced by other more indirect threats to personal well-being, including concerns about technology and risk. More and more people have attained a level of economic security that allows them to take up concerns beyond those of feeding and housing themselves and their families, securing basic health care, and providing for these security needs for their old age. And, regardless of socioeconomic level, people whose chief personal values extend beyond personal security are more likely to be concerned with environmental problems than the average citizen (Dunlap et al., 1983; Inglehart, 1977). Thus it is not surprising that affluence has brought increasing concern about the risks of technology.

Increasing Dependence of the Economy on Technology

The U.S. and world economies have come to depend increasingly on advanced technology for the production of food (petrochemicals), health care (drugs and other medical technologies), communication (computers and information transmission technology), transportation (jet aircraft), manufactured goods (automation and electric power technologies), and, of course, military security. Such technologies have increasingly been controlled by large, politically and economically powerful organizations with vested interests in discovering, developing, and implementing them. They are also supported by individuals who benefit from them economically or in other ways. The new technologies offer great benefits to their sponsors in money or political power and potential benefits and risks to society that may also be large—but poorly understood. The sponsoring organizations need public acquiescence to achieve their technological aims, but for the reasons discussed below that acquiescence has become more difficult to achieve. At the same time proposals to restrict technologies typically meet intense opposition from powerful proponents.

Distrust of Institutions

Public opinion polling data indicate that there has been a "sharp decline of public faith in government, business, and labor since the mid-1960s" (Lipset and Schneider, 1987:40). The decline was especially rapid between 1964 and 1975. Other polls have shown similar results, but the decline has been partially reversed more recently (Lipset and Schneider, 1987). The decline in trust in major institutions was in sharp contrast to the especially low level of criticism, distrust, and rebellion in the 1950s (Schudson, 1978). It was, no doubt, influenced by a series of formative political events of the 1960s and early 1970s. The civil rights movement, the war in Vietnam and the protest against it, the assassinations of three major national leaders, and, finally, the Watergate scandal all forced attentive people to look at the dark side of our national character and national institutions.[5] A climate developed in which major decisions by government and industry, including decisions about technology, were increasingly open to question.

The Environmental Movement

A social movement concerned with environmental protection developed in the 1960s in the United States and has since become a regular participant in technological debates. Influenced by new scientific knowledge conveyed in works like *Silent Spring* (Carson, 1962), large numbers of ordinary people saw for the first time that their personal interests or values were affected by the way society used and regulated technology. They expressed their concerns through environmental and related organizations and by direct pressure on government for action. Although environmental organizations were not new on the American scene, those that had existed before the 1960s, such as the Audubon Society, the Nature Conservancy, and the Sierra Club, had focused mainly on the conservation of wildlife and wilderness. The new organizations, and to some extent the old ones through changes in their political agendas, advanced a new brand of environmentalism concerned with threats to ecosystems and global and regional life-support systems and with the protection of people from technologically based threats to health and well-being (Hays, 1987). The new environmental organizations and their political allies gained widespread public support and raised funds to lobby, to conduct independent scientific analyses of technological issues, to participate in regulatory decision processes on matters of concern to their supporters, and to challenge government and corporate decisions in court. They have became an institutional presence in opposition to a range of efforts by industry and government to implement controversial new technologies and to further spread existing ones.[6]

New Public Institutions

During the 1960s and 1970s national institutions were being restructured to pay more attention to social goals, including improved management of societally shared risks. Beginning with passage of the National Environmental Protection Act in 1969, several new government bodies, such as the U.S. Environmental Protection Agency (1970), the Occupational Safety and Health Administration (1970), the Consumer Product Safety Commission (1972), the Nuclear Regulatory Commission (1975), the Office of Technology Assessment (1972), and the Office of Disease Prevention and Health Promotion (1984), were created to promote and protect public safety and health in specific areas of risk. Courts began to require that medical professionals provide patients with better information to guide their decisions about their treatment, and formal procedures for "informed consent" came into being (Applebaum et al., 1987; Faden and Beauchamp, 1986). Federal agencies, for their part, began to make more information about risk available to the public, for instance by requiring recordkeeping of the life histories of toxic substances. These changes created new public institutions whose purpose was to make technological decisions in the public arena and that resulted in new settings for conflict.

POLITICIZATION OF THE TECHNOLOGICAL DEBATE

The above changes in risks, knowledge, and society have contributed to the increasing conflict about technology in recent decades. The benefits of technology have increased, but many people believe the risks have as well. The hazards confront more people than ever before (even if the risks may be less), and they have gained the attention of a wider range of political actors. The attendant choices have huge potential effects on the distribution of wealth, health, and even political power in society. It is no wonder, then, that technological choices have come to concern more people and that the nature of those choices has come to be seen in a different light. As traditional political issues such as public health, social equity, and due process became more prominent in technological decision making, decisions that had been treated as essentially technical and economic, to be decided by executives of firms and government agencies with the advice of experts, came to be seen as also being essentially political (Dietz et al., 1989). The trend toward public involvement can be seen in a recent expansion of "right-to-know" legislation, the effect of which is to disseminate information that citizens can use to heighten their political involvement. The redefinition of environmental problems as political is evident in a number of changes in the political system, as described below.

Concepts of Regulation

Changes in federal law in the mid-1960s transformed the judicial concept of public interest as used in administrative law in regard to regulatory agencies. Regulatory proceedings were opened to more than just the parties who suffer direct legal injury from government action (*Office of Communication of the United Church of Christ* v. *Federal Communications Commission*, 1966; *Scenic Hudson Preservation Conference* v. *Federal Power Commission*, 1965). The New Deal notion of a regulatory agency as the embodiment of the public interest gave way to a concept of the regulatory agency as a political, quasi-legislative forum for the meeting of competing interests (Ackerman and Hassler, 1977). It is no wonder, then, that the EPA faced a rapid rise in the number of civil lawsuits challenging its regulations, from under 20 in 1973 to nearly 500 in 1978 (O'Brien and Marchand, 1982:80).

Tort Law

Tort law has changed, broadening the ability of different kinds of people and groups to bring legal action and creating new ways for plaintiffs to sue successfully even when there are formidable difficulties involved in determining who is responsible for an injury to the plaintiff. In the past 30 years private-law adjudication has moved away from caveat emptor and related rules to permit greater access to the judicial arena and to apply more flexible doctrines regarding compensation for environmentally caused damages to health and safety (O'Brien and Marchand, 1982). In the California Supreme Court decision in the case of *Sindell* v. *Abbott Laboratories*, for instance (a decision the U.S. Supreme Court let stand in 1980), the court allowed mothers whose children had suffered injury because of the mother's use of diethylstilbestrol (DES) to recover damages without being able to identify a particular manufacturer as responsible for the injury. The plaintiffs were allowed to recover by suing those manufacturers who collectively represented a major share of the market for the product that caused the injury (O'Brien and Marchand, 1982).

Regulatory Procedures

Regulatory rule making over the past two decades has evolved a set of procedures that guarantees a variety of interested parties the opportunity to comment on proposed rules and that makes it increasingly likely that regulators will have to address those comments as they justify their decisions (Schmandt, 1984). Federal agencies are required by the courts to prepare detailed scientific analyses in support of regulatory actions. These changes occurred in response to increasing conflict about risk and created a channel for the expression of opposition to government agencies' positions. They imposed some limits on what opponents could legitimately raise as objections, but at the same time the new procedures gave the opponents predictable access to the decision process and new opportunities to challenge decisions in court.

Politically Potent Symbolic Events

A number of incidents have received widespread attention and have become cognitive markers of danger for many people. Just as "Watergate" is synonymous for many with governmental malfeasance, so "Three Mile Island" has come to represent the dangers of high technology. "Bhopal," "Chernobyl," and "Love Canal" are other such symbols. These reach out beyond the immediate media coverage they receive to become part of the cultural consciousness of many people, even those who know little of or paid little attention to the original incidents (Slovic, 1987). As a result, the mere mention of these incidents can be a trigger for argument.

Increased Focus on Science in Technological Debates

The laws and procedures that control governmental decisions about technology in the United States have come increasingly to demand scientific and technical knowledge. Some regulations require government to determine whether a particular risk exists and to act accordingly; others require a determination of the "best available technology"; and others explicitly require a weighing of costs and benefits. The National Environmental Policy Act requires the preparation of careful assessments of the environmental and socioeconomic impacts of major technological choices. All these developments put science and scientific disagreements at the center of technological debates. Because of the difficulty, as discussed in Chapter 2, of gathering and interpreting all the scientific knowledge relevant to modern technological decisions, there is considerable room for scientists to disagree. When a decision that may have major political effects by altering the distribution of money, power, and well-being in society is made through procedures that emphasize scientific judgment, scientific disagreements tend to become proxies for political disagreements, and political adversaries often express their positions in the language of science (Dickson, 1984; Mazur, 1981; Nelkin, 1979a). In this way the inherent difficulty of understanding technological choices combines with the political importance of their effects to multiply the intensity of conflict.

Institutionalization of Scientific Conflict

Partly because regulatory decisions now rely so heavily on the evaluation of scientific knowledge, divisions in the scientific community have become increasingly public. Conflicts that might once have been contained within professional societies now appear occasionally as front-page news. Some environmental organizations and groups of scientists, such as the Federation of American Scientists, whose members share common concerns about controversial technologies, have built scientific resources that allow them to advocate political choices in the technical language of risk and benefit analysis that statutes and regulatory procedures often require. Not to be outdone, industry-based groups have increased their capability to do "regulatory science" in support of their positions on the same issues. Thus disagreements between scientists have gained an institutional place in the political debate, with scientists whose analyses support particular positions presenting their judgments on behalf of groups advocating those positions (Schmandt, 1984).

IMPLICATIONS OF CONFLICT FOR COMMUNICATION

The above discussion makes clear that many factors have contributed to increasing social conflict over hazards and risks. The conflict itself is a multifaceted one. A review of the environmental policy literature has identified four distinct aspects of risk conflicts, as described below. According to a recent survey of scientists, lawyers, and others whose careers are largely devoted to thinking, researching, and debating about technological choices, each of these is a major source of controversy about environmental risk (Dietz and Rycroft, 1987; Dietz et al., 1988).[7] This section distinguishes these four aspects of technological conflict and discusses the implications of each for risk communication.

Differential Knowledge

One source of conflict about risk is that experts and nonexperts know different things about the risks and benefits of technology. In particular, technical experts have specialized knowledge about the nature of both the hazards and their benefits that nonexperts, lacking this knowledge, may dispute. Conversely, nonexperts sometimes have local knowledge about exposures or the practical operation of a hazardous activity that technical experts do not share. When conflict arises mainly from differential knowledge, risk messages focused on information, which promote the sharing of knowledge, can improve the risk communication process. This realization underlies proposals to design messages that would explain to nonexperts in a clear and simple format what scientists and technologists know about particular risks. It also provides justification for the flow of informational messages from nonexperts to experts. In conflicts that arise from differential knowledge, better sharing of knowledge may also help reduce the conflict. However, when a conflict is in large part based on other factors, sharing of knowledge may not resolve it. It may even adversely affect the risk communication process if it is perceived as a diversion from the real issues.

A second aspect of differential knowledge and conflict is the differences in the degree of understanding in various groups typically involved in risk issues. Information simply made available to the public through the mass media and other channels is typically taken up more readily by those with high, rather than low, socioeconomic status because the former usually have a higher level of education, enabling them to understand technical material more easily. This leads to what is called a knowledge gap. But the presence of a conflict can change this situation. In certain circumstances the presence of conflict might be seen as positive because it effectively increases the number of people who become informed about the issues involved.

Vested Interests

Those who bear the risks of a technology are not always the same people who gain the benefits, and, when the risks and benefits are distributed in unequal proportion, those holding different interests come into conflict. This kind of conflict is most clearly evident in decisions about the siting of locally unwanted facilities such as hazardous waste sites, power lines, and radioactive waste repositories, but it is characteristic of other conflicts about risk as well. When a conflict is based in large part on vested interest, risk messages can be helpful if they clarify what different groups' interests are and describe how the available options would affect each of those interests. Such messages improve risk communication by providing information relevant to the choices at hand. But they often do not resolve conflict. Even messages that simply describe scientific information can exacerbate conflict if the information helps clarify who stands to win or lose.

Value Differences

Differences in values also underlie conflict about risk. For instance, some people may believe that a potential catastrophe should be avoided by not adopting a technology that might produce it, while others may believe that potential problems could be solved after the technology is implemented but before the problems become too serious. In trade-offs between economic growth and threats to health and to esthetic, ecological, or community values, political participants who expect the same outcome may still disagree with each other because what they may gain or lose does not have the same value to each of them. The source of such disputes may lie in people's relative preferences for values (e.g., money versus beauty), their beliefs in society's ability to control technologies once introduced, or their predispositions about how much risk to take under conditions of uncertainty. When a conflict is based in large part on differences in values, the following types of messages can make risk communication more successful: statements identifying the values at stake, arguments about which values deserve the most weight, and analyses of how each available option would affect different values. As with conflicts based on different interests, messages that improve knowledge relevant to the choices at hand and that therefore raise the quality of risk communication can at the same time make the conflict more intense. Even messages describing scientific analysis can have this effect, by clarifying which values an alternative would advance or impede.

Mistrust of Expert Knowledge as Interest Serving

Public mistrust of information from government and industry sources also underlies conflict about technology. Many people are aware that experts can be found who will support nearly any position in a technological debate. They realize that industry groups tend to produce only those scientific arguments that advance their goals and that environmental groups do the same. They know that even the federal government has been subject to strong accusations that its scientific analyses have been influenced by political pressure from various interest groups (e.g., Nelkin and Brown, 1984; Smith, 1983). Thus the statements of scientific experts in risk debates are seen by the skeptical parts of the public as reflecting political positions rather than unbiased assessments. Particular types of messages cannot by themselves alleviate mistrust, although altered procedures for the design of risk messages may help (see Chapters 6 and 7). Rather, the effect of mistrust is to make communication more difficult in all contexts.

Note for Risk Message Designers

In most risk debates some participants are concerned with narrower issues of risk analysis, some with interests, some with value questions, and some with issues of trust. For this reason, different participants want to send and receive different kinds of risk messages, and the risk communication process includes the full range of types of messages mentioned here—scientific analyses, expressions of interest and value, and arguments about which values to favor. The designers of risk messages need to be aware that a program of messages that addresses one source of conflict may fail to address other sources. Thus someone who designs a message to eliminate differential knowledge may find an audience concerned with interests or values or one that mistrusts the message source—and the message may not have the desired effect. Such a message may even intensify conflict because the audience sees it as irrelevant or as a diversion from what it considers to be the main issue.

Risk communication is difficult in part because risk messages often seem to operate at cross-purposes. The next chapter distinguishes the major settings of risk communication and the major purposes for risk messages. It explores the issue of what techniques are appropriate for risk messages, particularly when the purpose is to influence the recipients' beliefs or actions.

NOTES

1. Conflict also occurs about the benefits of technological choices. This chapter discusses the risks because they have usually been the focus of the most intense conflict.

2. The headings "It is the safest of times" and "It is the riskiest of times" are quoted from Denton Morrison's paper, "A Tale of Two Toxicities" (1987).

3. Although public support for increased environmental regulation is strong, as evidenced by direct questions on opinion surveys, environmental problems are not usually mentioned with great frequency in response to open-ended questions such as, "What are the three most important problems facing the nation?"

4. Some types of cancer are clearly linked to chemical exposures: mesothelioma and asbestos, vaginal cancer and diethylstilbestrol (DES), bladder cancer and benzidine dyes. In these situations the inference about possible causal agents involves assessment of statistical evidence (e.g., epidemiological studies) and biological evidence on the plausibility of the linkage between agent and disease [e.g., gasoline vapors cause kidney tumors in male rats, but the mechanism is not believed applicable to human kidney cancer (EPA Science Advisory Board, 1988)].

5. Research on the ways social movements mobilize citizens' attention and participation has recently been reviewed by Cohen (1985) and Jenkins (1983).

6. Recent studies on the growth of the environmental movement include those by Hays (1987), Milbrath (1984), and Touraine et al. (1983).

7. That is, each of these four aspects of conflict was rated as a major source of controversy about environmental risk by a majority of the "risk professionals" in the survey sample.

4

Purposes of Risk Communication
and Risk Messages

In this chapter we distinguish two types of settings—public debate and personal action—in which risk decisions and risk communication occur, and we show how the risk communication process and its participants vary in these settings. We then discuss two distinct purposes of risk messages—informing and influencing—that coexist in risk communication, sometimes even in a single risk message. Finally, we address the thorny ethical problem of the appropriateness of influencing as a purpose of risk messages, particularly messages that public agencies distribute to citizens.

SETTINGS OF RISK COMMUNICATION

Public Debate

In a setting of public debate—such as congressional hearings, congressional debates, formal regulatory adjudication, and notice-and-comment rule making—democratic risk communication includes a wide range of messages, sources, and audiences. Interested groups raise questions for the experts, who respond; experts from different perspectives dispute with each other; and citizens and their representatives dispute using, among other things, the experts' findings and criticisms of each other's results. Messages describing and summarizing scientific knowledge about risks and benefits are important, as are critiques of those messages and that knowledge. In the United States, regulatory decisions must generally be based on the best available scientific knowledge to be defensible against legal challenges. As a result, much risk communication in the regulatory context deals with the adequacy and proper interpretation of scientific evidence. But risk communication also includes expressions of opinion, concern, frustration, and the like by all participants, directed at whomever will hear and might act. Such decision making tends to be adversarial, with political actors making the strongest possible case for their positions, overtly expressing their interests and values or citing expert judgment and analysis depending on which arguments seem most effective. Recipients of risk messages understand that those messages are guided by interests and political positions and so do not expect any single source to offer an unbiased assessment of available scientific knowledge.

Public policy about tobacco smoking illustrates the range of risk messages that come out of public debate. The policy options for risk management involve decisions to be made in different bodies, each using different rules of debate and assigning different roles to the general public within those rules. For instance, the federal government has considered increasing excise taxes on cigarettes, placing warning labels on cigarette packages, funding antismoking advertising campaigns, distributing informational pamphlets on the health hazards of smoking, and banning smoking in various public places. Other options that might be considered for cigarettes, and that have been used for other health hazards, include outright prohibition on manufacture or sale and restriction to use by prescription only. In state and local governments, debates have also proceeded on options such as banning cigarette advertisements in some public places, raising the minimum age for purchasing tobacco products, banning smoking in municipal buildings, and requiring no-smoking sections in restaurants.

Risk communication varies from one of these decision-making arenas to another. Citizens participate in legislative settings by attempting to influence their representatives directly or by affecting the general climate of opinion and thus achieving indirect influence. In federal regulatory decision making, there is also wide latitude for participation, although the Administrative Procedures Act and agencies' practices constrain the time and type of participation and the kinds of arguments that can be introduced (Greenwood, 1984). Agency procedures differ, particularly in terms of how much two-way communication they allow and how much they do to provide expert knowledge to the citizenry at large. Nevertheless, public debate in the regulatory or legislative context allows for risk messages and other related messages from a large number of sources.

We consider risk communication in a setting of public debate successful to the extent that it raises the level of understanding of relevant issues or actions among the affected and interested parties and those involved are satisfied that they are adequately informed within the limits of available knowledge. As noted in Chapter 1, successful risk communication does not imply optimal risk decisions; it only ensures that the decisions are informed by the best available knowledge. Also as noted in Chapter 1, raising the level of understanding requires more than making accurate information accessible to the interested parties. Success requires increased understanding of the issues to the extent that the parties involved desire to understand. Although individual risk messages may contribute to increased understanding, the net effect of risk communication on understanding depends on all the messages individuals receive and their interpretation of them. Therefore, the designers of risk messages who wish to increase the recipients' understanding need to take into account the recipients' willingness and ability to receive and understand the messages as well as the effects of other, sometimes conflicting, messages that they may also receive.

Success for risk communication does not require that every citizen be informed about the risks presented in every regulatory decision, but people need to be confident that some person or group that shares their interests and values is well informed and is representing those positions competently in the political system. Public debate, in a traditional view in the United States, implies a pluralism of constituencies, with "consent of the governed" consisting of trust that the relevant views are represented, that the procedures do not disadvantage important constituencies, and that the people are able to hold public officials accountable for their actions.

The requirement that interested parties believe they are adequately informed is worth explanation. It stems from recognition that in several arenas of public debate risk decisions are intensely controversial and many message sources are widely mistrusted. This situation imposes requirements, particularly on those message sources and in those policy arenas, that may seem unfair to officials who believe their responsibility to the public extends only to making wise decisions and providing complete, accurate information. But if a message source is widely mistrusted, its messages will be rejected by many regardless of completeness or accuracy. If accurate information is rejected by recipients, it does nothing to increase their knowledge base—hence the requirement that recipients of information for public debate be satisfied that they are adequately informed.

Both of the above-mentioned requirements for successful risk communication were factors in the public debate that resulted in the successful siting of the ECOFLO hazardous waste facility in Greensboro, North Carolina. This siting case also illustrates an instance in which understandable and sensitive messages from an individual risk communicator (ECOFLO) contributed to the success of the overall risk communication process involving the Guilford County Hazardous Waste Task Force, environmentalists, and other concerned citizens (see accompanying story, pages 76–77). Nevertheless, it should be emphasized that open and free communication will not necessarily ease conflicts in all situations.

With respect to a designated decision maker, such as the head of a regulatory agency, risk communication is successful only if it adequately informs the decision maker. A decision maker is adequately informed within the limits of available knowledge if provision of all remaining available information would add nothing to justify a modification of his or her choice. Decision makers need to be informed about the managerial and political aspects of the choice at hand as well as about the state of technical knowledge. And, as already noted, the relevant knowledge should be understood by the decision maker, not merely made accessible.

It is important to emphasize that *a successful risk communication process is different from a risk message that is successful from the standpoint of its source.* In a public debate (like that in the ECOFLO case), participants produce risk messages aimed at changing minds and influencing political outcomes. From their perspective a risk message is successful to the extent that it contributes to the outcomes its sponsor desires. Sometimes a risk communicator will make false or deceptive statements or will withhold pertinent information to achieve a political effect. Such activities, if they are not revealed, may achieve the ends of the message source but not the social goal of an adequately informed debate.

ECOFLO HAZARDOUS WASTE FACILITY SITING
GREENSBORO, NORTH CAROLINA

The successful siting of the ECOFLO hazardous waste facility in Greensboro, North Carolina, in 1985 is an example of good risk communication and effective risk messages. Although representing a situation somewhat less problematic than those encountered elsewhere—the company proposed a treatment facility to reduce the overall amount of toxic material in that locale—it does illustrate the role of communication efforts in the siting of a hazardous waste facility. The siting of such plants is notoriously difficult. As a result of ECOFLO's efforts, however, the final public hearing to site the facility lasted only 15 minutes and led to the permitting of the plant with the blessing of local government officials and environmentalists (Lynn, 1987).

ECOFLO began operation in Greensboro in September 1983 with a license as a waste transporter. It worked mainly with small companies that produced about 20 drums of waste a month. Although ECOFLO was a new company, its owners had previously worked for other hazardous waste companies. In July 1984, ECOFLO submitted its plans for a hazardous waste treatment facility to the state of North Carolina. The plant was designed to serve primarily local and intrastate markets and would not handle PCBs, dioxins, cyanide, radioactives, biological wastes, or explosives. The treatment processes to be used were neutralization and centrifugation. Wastes that had to be burned would be transported elsewhere (Lynn, 1987).

A year and a half prior to ECOFLO's application, another company had tried to site a hazardous waste facility in Greensboro and failed. Local citizens, unable to receive information or to have their concerns addressed, had successfully organized opposition to that facility.

The Greensboro area had a group of citizens well versed in hazardous waste issues. As a result of an EPA grant to the North Carolina League of Women Voters in 1979, the Guilford County Hazardous Waste Task Force was formed. The task force sponsored short courses on toxic materials and workshops and displays to educate and organize the community. By 1985 the task force and its chair, Carolyn Allen, had good working relationships with the local government staff and elected officials.

When ECOFLO decided to site in Greensboro, the task force invited neighborhood leaders from the part of the city where the facility might be located to a series of education meetings on hazardous waste. These workshops included the chemistry of hazardous waste, disposal processes, and a session with Tom Barbee, ECOFLO's vice president and the Greensboro plant manager (Lynn, 1987).

This was not Barbee's first contact with the task force. He had been attending task force meetings since 1979, as a professional waste manager with another firm. He was also a native of North Carolina and a longtime Greensboro resident. He did not see the environmentalists as the enemy. In a local TV interview he said that ECOFLO "honestly wants to be a service to the community. . . . We want to help local companies handle their waste as responsibly as possible. . . . We are on the side of the environmentalists" (quoted in Lynn, 1987).

From the time ECOFLO decided to site a facility in Greensboro, Barbee had been contacting relevant groups and individuals. He went to the local police and fire departments to ask what they thought he needed to do to ensure a safe site. He talked with ministers, neighbors, the planning and zoning department, and county commissioners. He gave candid and detailed answers to questions by citizens. He and his staff took the press, state and local officials, and neighbors on plant tours. He even sponsored his own public meeting before the state held its public hearing. Barbee's meeting was cohosted by Bruce Banks, a local chemistry professor and Audubon Society member; Carolyn Allen, chair of the task force; and Jim Rayburn, chair of the Guilford County Advisory Board on Environmental Affairs (Lynn, 1987). At this meeting Barbee detailed how he had made changes in his original proposal based on feedback from the fire department, the planning commission, and the task force, among others. He invited public participation and took the public's concerns and suggestions into consideration in ECOFLO's revised plan.

This willingness on the part of ECOFLO to involve the community, to share information, and to implement changes based on community input proved effective. The ECOFLO waste treatment facility was approved and the citizens were satisfied it could be operated safely (Lynn, 1987).

Personal Action

Risk communication regarding personal action is quite different from risk communication regarding public decisions. At minimum the setting is more limited because most risk messages are addressed to individuals rather than to a spectrum of participants in public debate. Sending messages to an individual is in one respect like sending them to the head of a regulatory agency: both have the ultimate authority to act. But the two situations are also different in important respects: few individuals have staffs of experts paid to answer their questions, and individuals seldom want the amount of detail that is justified when a federal regulator is about to make a decision for the whole population (see Figure 4.1). Much of risk communication in this setting takes the form of messages directed at the public offering information, advice, warnings, or recommendations regarding risky individual actions. Both public agencies and private organizations sometimes design such risk messages. But personal action is also influenced by a variety of risk messages, usually informal, from other individuals. People want to know how hard it was to stop smoking, or whether low-fat meals can be made to taste good, or in what ways other people feel better after losing weight. Such risk-related messages, regardless of whether they accurately represent the likely outcomes of alternative actions, can be critical in individual decisions (Nisbett and Ross, 1980).

Tobacco smoking also illustrates the kinds of risk communication issues that arise in the context of personal choice. Despite the restrictions created by recent policies, people still choose whether, how much, when, and where, within limits, to smoke. But Congress has decided that it is in the public interest to influence smoking behavior in various ways short of directly restricting tobacco use. Cigarette taxes and advertising restrictions are two policies that constrain individuals and the tobacco trade. Other policies, such as the requirement of warning labels and widespread dissemination of the surgeon general's findings on the risks of smoking, rely on risk messages as an alternative to direct control of the substance. Such policies create a risk communication setting much different from that of public decision making, particularly because they call for specialized risk messages. Congress has sanctioned efforts by government officials, including the surgeon general and other medical experts, to design and disseminate messages aimed at changing individual behavior.

We consider risk communication in the setting of personal choice

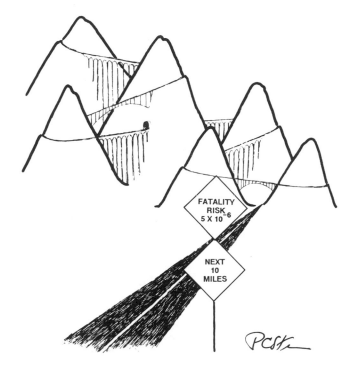

FATALITY
RISK
5×10^{-6}

NEXT
10
MILES

FIGURE 4.1 For personal action to reduce risks, a simple warning sign (e.g., "Hills and Curves Next 10 Miles") may be sufficient; a report of a formal risk analysis could be counterproductive. SOURCE: Courtesy of Paul Stern.

successful only if it adequately informs the individual for making a choice among alternatives. Adequate information, to reiterate, must be understandable for risk communication to succeed; it is not sufficient that it be available. Part of the debate is about going further, so that the recipients are somehow brought to understand the material. But we have not gone so far as to include this as a criterion for success.

Getting recipients' attention and comprehension poses significant barriers to risk communication, especially in the arena of personal action, where many recipients customarily act without carefully considering risks and benefits. It should be noted that from the standpoint of the designers of risk messages, the goal may or may not be to inform choice. Often a message is intended to influence choice, a very different matter, even if experts believe that the choice they desire to elicit is in the audience member's interest. Thus some risk messages from government agencies are designed to inform choice (e.g., nutritional information on food packages), but at other times, occasionally after open debate in a legislative setting, an explicit decision is made to influence beliefs or behavior in a particular direction (e.g., anti-drunk-driving campaigns). Although risk messages are sometimes judged against a criterion of behavior change, this is not an appropriate test of whether an individual has made an informed choice. It is possible for an individual, fully informed of the risks, to choose to engage in hazardous behaviors such as smoking, skydiving, or leaving seat belts unbuckled.

Sometimes risk messages are intended to inform or explain rather than to be used as direct input to a choice. This can be the case when the risk manager is in the position of explaining a decision that has already been made. It can also occur in situations when individuals or groups are unavoidably exposed to particular hazards. It may be necessary to explain why a decision has been made that is injurious to the recipients of the message or that has other undesirable consequences.

INFORMATION AND INFLUENCE: THE PURPOSES OF RISK MESSAGES

We have noted that successful risk communication, such as that described in the ECOFLO case, makes for better-informed decision makers, both individuals and public or private officials. A "successful" risk message, in contrast, is not always one that increases the understanding of decision makers. For risk messages success is commonly interpreted in relation to the goals or purposes of the message source. The sources of risk messages sometimes aim to inform the recipients, but sometimes they aim to influence their beliefs or actions. A risk message designed to influence may be judged successful even if it does nothing to add to the audience's understanding. An antidrug campaign that relies on exhortations from prominent sports figures is successful if it keeps some teenagers from addiction, even if they learn nothing new about the health effects of heroin or cocaine.

We recognize that efforts to influence through risk messages do not always have such noble purposes. The sources of risk messages may set their own criteria of success but attaining them does not always advance a public good. Sometimes "effective" risk messages are inconsistent with promoting substantive public good, as when they mislead people about what is in their interest. At such times they are in conflict with the public goal of successful risk communication. (Sometimes, however, audience members gain understanding even from biased risk messages. For instance, judges, elected officials, and interested citizens often gain understanding on matters of public controversy by comparing messages from various sources that they realize are trying to influence them. They inform themselves, despite the efforts of message sources to influence rather than inform.)

Serious confusion can arise because any given risk message may be intended to inform or to influence. It can be difficult for a recipient to tell which aim a particular message has; message sources, aware of this difficulty, sometimes attempt to persuade in the guise of informing. That tactic is likely to be most effective when it goes undetected,[1] but it can backfire seriously if revealed, undermining the credibility of the message source and creating resentment and mistrust. The problem of dual purposes is compounded by the fact that the designers of risk messages are often called on to both inform and influence the same audience with the same message. Regulatory agency employees, for instance, are routinely asked to prepare a document to support a decision at the end of a formal rule-making process that both summarizes the evidence on which the decision was based (thus informing the audience) and justifies that decision (thus endeavoring to influence the audience to believe the right choice has been made).

The dual purposes of risk messages complicate defining responsible behavior for the designers of the messages. In order to arrive at some criteria for the acceptability of attempts to influence, we begin by describing a dimension along which one can array techniques for the construction of risk messages. At one end of the dimension is an ideal, pure information, free of techniques of influence; at the other end is deception. Although the purpose of informing is consistent with the goal of successful risk communication—to raise decision makers' level of understanding—the use of techniques that aim to persuade, deceive, or otherwise influence decision makers implies that a different goal is being pursued.

Information

To inform someone about an issue or choice is to assist that person to apprehend the relevant propositions or statements that describe the issue or choice. Ideally, the result is that the person or persons informed gain a full or complete understanding of the issue or choice. This appears to have happened in the ECOFLO case. In practice, however, full understanding does not exist for most important choices about risk (see Chapter 2), so it cannot be conveyed. *A practical goal for information is for the recipient to gain understanding, within the limits of available knowledge, that is adequate to make appropriate choices given his or her values.* Adequate understanding does not require knowing everything that is known about an issue, only enough to be able to make choices in one's own best interest. If more precise information would enable members of the audience to make choices that better approximate their desires, it should be provided; if it would not aid in decision making, more precision is unnecessary.

Influence

A spectrum of techniques is available for designing risk messages that go beyond pure information and that can be used to influence an audience. The most extreme techniques involve outright deception: strategies such as "lying, withholding of information, true assertion that omits a vital qualification, and misleading exaggeration to cause persons to believe what is false" (Faden and Beauchamp, 1986:363). But many influence techniques do not do such violence to the truth. In order to consider the appropriateness of different techniques, it is useful to identify them. The following paragraphs describe different techniques, beginning with some that stay close to the facts and moving to some that do not depend much on factual information. Some of these techniques can be used either to inform or to influence. It is this possibility that makes it difficult for recipients of risk messages to determine their intent and therefore to interpret their content.

Highlighting Facts

Risk messages cannot include all the details known to science and still be read and understood by most nonexperts. Therefore the designers of messages omit some information and highlight other information. For instance, message designers choose whether to summarize knowledge about both possible deaths and illnesses arising from a risk or only about deaths, about both direct and synergistic effects or only direct effects, about effects on subpopulations including sensitive groups or just on whole populations, and so forth. Having chosen what to present, message designers must also make choices about what parts of the message to emphasize with visual aids, vocal emphasis, underlining, color, and other techniques. Although highlighting may be employed only to emphasize the essentials of what is known, decisions to highlight—which are unavoidable—involve judgments about what is essential. A large psychological literature demonstrates that highlighting information, or making it more "available," affects the understanding and the decisions of those who receive the messages (Fiske and Taylor, 1984; Kahneman et al., 1982; Tversky and Kahneman, 1973). Thus highlighting can influence the audience's beliefs about what aspects of a risk decision are important in the direction desired by the message designer.

"Framing" Information and Decisions

Different ways of presenting the same facts can create different impressions. When a risk estimate is uncertain, it can be described by a point or "maximum likelihood" estimate or by a range of possibilities around the point estimate. But estimates that include a wide range of uncertainties can imply that a disastrous consequence is "possible," even when expert opinion is unanimous that the likelihood of disaster is extremely small. The amount of uncertainty to present is a judgment that can potentially influence a recipient's judgment.

Another example of "framing" involves the choice between alternative ways of presenting the same numerical information. One study, for example, found that a hypothetical vaccine that reduces the probability of contracting a disease from 0.20 to 0.10 is less attractive if it is described as effective in half the cases than if it is presented as fully effective against one of two virus strains that strike with equal probability and that produce the same disease. This finding suggests that people favor full protection against an identified risk over equivalent but probabilistic protection (Tversky and Kahneman, 1981). Similar differences in presentation have been identified with respect to whether outcomes are presented in terms of "sure loss" or an "insurance premium" (Fischhoff et al., 1980) or "lives lost" as opposed to "lives saved" (Tversky and Kahneman, 1981). It has even been demonstrated that when two versions are presented sequentially people often reverse their preference from the first presentation to the second (Hershey and Shoemaker, 1980).

Risk Comparisons

An important instance of framing is the use of risk comparisons. Comparing one risk that is not well understood to another that the audience comprehends may be a useful way to convey information about the former risk. It is often difficult, however, to find risks that are similar on enough attributes to carry the comparison. But risk comparisons can also be used to influence or even mislead, because a risk comparison may improperly carry the implication that if a person is willing to take the larger of two risks he or she should accept the smaller as well (Covello et al., 1988; Fischhoff et al., 1981a). The uses of risk comparisons are discussed in more detail in Chapter 5.

Persuasive Use of Facts

Risk messages often involve a selection of the facts to make a point. Messages aimed at convincing recipients of a point of view can use techniques of highlighting and framing but can also employ other rhetorical techniques: selective presentation of evidence, creation and destruction of "straw-man" arguments, judicious placement of the various arguments within a message for maximum effect, listing of supporting arguments by number to make the argument look stronger, and so forth. Such techniques can enhance the persuasive effect of messages, sometimes without any alteration of the content (Cialdini, 1984; Eagly and Chaiken, 1985; McGuire, 1985), and they can be quite difficult for a recipient to detect.

Appeals to Authority

Nonexperts often want to know who has taken what position on a difficult choice before them. When they do not know enough to make an informed choice themselves, or believe it too expensive or time consuming to become fully informed, they may choose to adopt the position of a person or organization they consider expert and trustworthy. Thus risk messages can be influential by supplying information about who has taken positions on an issue. They may be balanced in their references to authority or they may not: a message may quote some scientists in support of a position but omit quotations from similar scientists who disagree. They may quote relevant authorities who have specialized knowledge or they may refer to sources widely trusted on other issues but ill informed on the issue at hand. And they may be accurate or inaccurate in representing the views of the authorities. Clearly, appeals to authority can fall at many different points along the dimension from pure information to deception.

Appeals to Emotion

Risk messages sometimes appeal to fear, pride, guilt, community spirit, parental concerns, or other emotions to spur people to action. Sometimes emotional appeals are made in the context of a presentation of information. Thus, saying that cigarette smoking causes emphysema conveys the same information with or without an accompanying film of an end-stage emphysema patient, but with the film the message will have a different effect. Appeals to emotion are not always more effective in inducing behavior change than less emotional appeals: the psychological research shows that the effect depends on other aspects of the message as well (Petty et al., 1988). Nevertheless, appeals to emotion can be effective influence techniques under some conditions. Sometimes the use of emotional appeals is widely accepted, but often it is considered manipulative and irresponsible. The conditions under which emotional appeals are considered acceptable are not well understood.

USE OF INFLUENCE TECHNIQUES IN RISK COMMUNICATION

Achieving Balance

Risk messages often employ some of the above influence techniques; indeed, it is difficult to imagine a risk message that could attract the attention of nonexperts without making use of at least highlighting or framing. A paradox arises for risk communication: How can messages be made to improve the recipients' base of information if, in order to be effective, they must use techniques of influence? The paradox disappears when one realizes that there are strategies for controlling the use of influence techniques consistent with the goal of successful risk communication. Substantive guidelines should be established for the content of risk messages that responsible message designers, including government officials, can to keep influence techniques under control so as not to bias recipients' understanding. Because available knowledge is inadequate to provide highly detailed substantive guidelines, procedural approaches that keep message designers in bounds are also critical to achieving successful risk communication.

The strategy of substantive guidelines is highly demanding. As already noted, the language of risk messages and even the measures used in risk analysis often embody value judgments or otherwise tend to lead the recipients of messages toward particular conclusions. We have noted several examples, but not enough is known to identify all the ways a risk message might bias a recipient's understanding. Thus it is not now possible to devise a complete guide to sources of potential bias that would allow risk messages to be evaluated for balance. Moreover, research on communication strongly suggests that the most effective message design for any particular purpose varies with the subject matter at hand, the decision alternatives, the intended audience, and other factors. But very little is known about the key situational variables that alter the effects of risk messages. Thus at present any guidelines for balanced risk messages would lack situational specificity. Existing knowledge can help message designers by identifying some potential pitfalls, but it cannot yield highly specific guidance. Responsible message designers need to interpret available advice, keeping in mind that knowledge is incomplete and that general principles may not apply to certain specific situations. Since there is no clear best way to make such judgments, substantive guidelines are not enough to ensure balance in risk messages, even when the sources are doing their best to achieve it.

The procedural strategy, which relies on a system of checks and balances to control the possible biases in risk messages, is applicable without regard to the state of knowledge about the effects of risk messages. The strategy assumes that available guidelines will never be perfectly correct or clear-cut and that vested interests or strongly held values will often induce ingenious message designers to find ways around guidelines. It therefore relies on systems of scrutiny and criticism, and the discipline of competing messages, to keep message designers within bounds.

Examples of procedural strategies applied to individual messages are the procedures of the National Center for Toxicological Research (NCTR) Consensus Workshops and those of the National Research Council (NRC) for review of its reports. The NCTR Consensus Workshop Series involves scientists from academia, government, industry, and public interest groups gathered to resolve toxicological issues, usually concerning the hazard posed by particular substances (Gough et al., 1984). Consensus is sought, not by formal voting, but through the chairman's guiding discussion toward agreement. Careful procedures ensure that all panelists have an opportunity to submit statements and to evaluate and comment on reports. These procedures ensure that reports focus on those areas where consensus is reached and present the major factors in reaching agreement. The NRC, many of whose reports are detailed messages about risk, does not rely on guidelines for the use of language, graphics, and so forth. Rather it relies on a balanced choice of committee members and an independent review process. The NRC presumes that a dialogue of well-informed individuals with varying perspectives will yield a first approximation of a balanced assessment. The outcome of this process is double-checked by submitting it to an independent review process involving experts who also represent a range of perspectives. In these two procedures it is not substantive guidelines but the process of dialogue and criticism that is used to ensure a balanced message.

Achieving Influence

Even more difficult than the problem of achieving balance in risk communication is the problem of deciding whether balance is the wrong objective. Advocates whose clear purpose is to influence their audiences may experience no problem, but the issue can be particularly acute for public officials who sit in a relation of public trust to the recipients of their messages. When should messages aim at merely informing the public, or government decision makers, and when should the goal be to influence the recipients?

Government officials are commonly expected to follow a more restricted standard of behavior in the area of risk communication than are advocacy groups, private citizens, or corporations. Similarly, citizens apply a stricter standard to messages paid for with public funds than to privately funded messages. We judge that such standards are justified because government officials hold a public trust. But the specifics of such standards are not easily defined.

After considerable debate focusing on the appropriate use of risk messages by public officials, we concluded that no explicit guidelines can be drawn defining which techniques are appropriate or inappropriate in particular situations or for particular message sources. We agreed that *informing is always an appropriate goal in the design of risk messages and that deception is never appropriate.* But we recognize that messages that employ influence techniques or that have influence as an objective are often considered acceptable, even coming from public officials. We believe that more extensive public debate is needed to arrive at standards for responsible behavior by public officials in the design of risk messages. As a contribution to that debate, we offer the following observations about the conditions under which influence techniques seem most likely to be considered appropriate by various audiences.

First, the acceptability of influence as a purpose of risk messages seems to depend in part on which beliefs or actions are being influenced. Consider the range of actions and opinions that government agencies have tried or might try to influence with risk messages. Here are some examples:

- Inoculating children against diphtheria, polio, pertussis, or swine influenza;
- Using condoms to prevent AIDS, gonorrhea, or pregnancy;
- Avoiding or reducing consumption of heroin, alcohol by drivers, tobacco products, alcohol by pregnant women, aspirin by children, or animal fat;
- Using seat belts, motorcycle helmets, or masks for painting or working with fiberglass;
- Supporting drug enforcement activities, AIDS research, EPA enforcement activities, or the repeal (or passage) of particular pieces of legislation.

Depending on the action or opinion in question, the likely response to government-sponsored influence attempts may vary from general acceptance to extreme controversy. Within each of the categories just listed, we believe that efforts to influence the action or opinion mentioned first would be relatively uncontroversial compared with similar efforts to influence the actions given later in each category. It is important to recognize, however, that observers, including members of our study committee, differ on the appropriateness of influence techniques in certain of the contexts listed. Some variation in judgments concerns scientific knowledge: the more clearly it has been established that an activity is dangerous or that it may harm persons generally considered to deserve societal protection (e.g., children), the more acceptable influence attempts seem to become.[2] But because of scientific uncertainty, informed observers sometimes disagree about how well established the relevant knowledge is. Another central issue seems to be the compatibility of governmental influence with individual autonomy and related values (Faden, 1987; Faden and Beauchamp, 1986). When a class of personal action (such as drunk driving) affects a large portion of the populace or threatens to inflict substantial monetary and other costs on society or on individuals who do not engage in that action, people are more willing to accept, and even to demand, that government agencies be proactive and try to influence beliefs and actions. Under such conditions, people are more willing to compromise the autonomy, privacy, or freedom of some individuals for the good of others.

Second, the acceptability of influence seems to depend on the techniques employed. Generally, the farther an influence technique lies along the dimension from information to deception, the harder the message becomes to justify and the clearer and more explicit must be the legitimate public purpose being served. To influence people to use condoms to prevent AIDS, government might appeal to authorities (the surgeon general recommending use of condoms to avoid AIDS) or respected or admired individuals (film and popular music stars hosting a TV special encouraging use of condoms in AIDS prevention), post warning signs (in lavatories of establishments frequented by homosexual males), present selected risk and risk reduction information ("use of condoms can reduce the transmission of AIDS by 95 percent"), or appeal to emotion (photographically depict the late stages of AIDS or state that "you sleep with your partner's whole sexual history"). Observers differ on the appropriateness of such techniques for a particular purpose, even when all agree that the purpose justifies some form of governmental influence.

We conclude that public values about the importance of particular public purposes and the acceptability of particular influence techniques are not well understood. Generally, the more an influence attempt would compromise important values such as personal autonomy or constitutional guarantees such as freedom of speech or association, and the more closely the influence technique approaches deception, the more it needs to be legitimated in order to be acceptable. Legitimacy is what makes people consider a particular influence attempt either responsible or irresponsible and either appropriate or inappropriate for government officials.

But there are no clear a priori guidelines that can tell a government official or other designer of a risk message when the message's purposes are sufficiently legitimate to justify a particular technique that goes beyond informing. Government officials will likely find their efforts to influence contested if they stray from accepted scientific views or if they challenge popular consensus. It is for this reason that decisions about governmental use of influence techniques in risk messages are often debated in overtly political arenas rather than being left to unelected officials' unscrutinized discretion. We believe that political arenas are the proper place for deciding the appropriateness of governmental efforts to influence citizens. *Governmental attempts to influence citizens' beliefs and actions can be justified only to the extent that some legitimate public process has culminated in a decision that using risk messages to influence behavior serves an important public purpose.*

Influence and Personal Action

The clearest example of politically established legitimacy for risk messages occurred in the congressional debate on persuading people to stop smoking. A congressional act codified language—a set of risk messages—that now appears on cigarette packages. The process of debate and approval by elected officials granted legitimacy to the messages.

Such explicit public debate rarely occurs to give clear prior justification for governmental attempts to influence personal behavior. Nevertheless, an agency or official can sometimes act legitimately on general authority. For example, public health officials have fairly general support in the mandates of their agencies for influencing people to take action to prevent the spread of infectious diseases. As a result, the surgeon general's 1988 mass mailing of a risk message about avoiding AIDS was met with wide public acceptance and even gratitude. Sometimes executive branch officials justify influence attempts within the spirit of their legislative mandates. The U.S. Environmental Protection Agency's efforts to inform the public about the health risks of indoor radon, and to convince people to have their homes tested and sometimes modified at considerable expense, are not justified by anything stronger than the EPA's general mandate for environmental protection. Yet this attempt to influence behavior in the setting of personal action was widely welcomed.

Influence and Public Debate

Sometimes executive branch officials rely on their general mandate to influence beliefs in the setting of public debate. Such efforts tend to be more acceptable after a risk management decision than before (e.g., when regulators are expected to justify their decisions to the public). But even before a decision is made, there are situations in which some kinds of efforts to influence public debate are appropriate. Regulatory officials sometimes argue that they have an obligation to evaluate new risks and, when public action is needed, to persuade elected officials of that fact. It is not enough, they say, merely to inform the public of the latest knowledge. Thus some public officials, on receiving evidence on the risk to the earth's ozone layer from chlorofluorocarbons, attempted to influence the highest levels of government to support an international treaty to cut production of that class of chemicals.

But it is easy for a public official to overstep the bounds of acceptability. This happens most readily when the subject matter of the influence attempt is already politically controversial or when government can be seen as trying to influence free political expression. When the San Francisco office of the Energy Research and Development Administration distributed 78,000 pamphlets defending the safety of the nuclear power industry during a 1976 California referendum campaign on the future of nuclear power, the result was a critical report from the General Accounting Office and strong expressions of congressional outrage (Burnham, 1976). Not only was the message unacceptable, but its dissemination and the agency's evasive response to criticism harmed the agency's credibility. With many influence attempts it takes fairly explicit debate and agreement to make them legitimate: vague appeals to an agency's mandate are not sufficient.

The judgment of whether public officials have or have not exceeded their proper role in a particular attempt to influence public debate is difficult to make. But it is a matter of judgment. Clearly, the freedom of public servants to influence decision makers must be kept within bounds. We considered and rejected the position that advocacy is always inappropriate for executive branch officials in the setting of public debate. There are situations in which such officials are in the best position to alert the public to a hazard that may deserve governmental action. But it is difficult to define the proper limit. Scientific analysis is indispensable to successful risk communication. It can show what is known about risks and the attendant choices and can identify the limits and uncertainties of that knowledge; it can therefore indicate what can be said. Science can also advise on when and how best to say it in order to improve an audience's understanding or to influence beliefs and actions. A decision to engage in advocacy, however, involves judgments about which risk management option is appropriate and about how much to influence audiences with other than information—judgments that must be based on values as well as knowledge. We concluded that natural and social sciences cannot provide guidelines for when to engage in advocacy in risk communication. Although empirical research can determine which beliefs Americans consider acceptable for influence by government and which influence techniques they consider most extreme and therefore most in need of legitimation, there is no practical way to tell in advance whether enough legitimation exists in the political system to justify a particular attempt to use risk messages to influence recipients. Advocacy messages from executive branch officials must therefore be judged against the legitimate role of the officials in question, as set forth in the relevant legislation and judicial interpretations and as argued by elected officials. *The decision of what are legitimate bounds for governmental risk messages is and ought to be made through the political process.*

We recognize that the boundaries for advocacy in the political process often are clear only after a public official has overstepped them, leaving public officials in an unpleasant position. However, such boundaries usually can be discerned in advance by careful analysis. In any case, when officials judge that the public welfare depends on a specific change in policy or individual behavior, they must also judge how far they can go before overstepping legitimate constraints. Advocacy can be politically risky for public officials. It may be widely applauded or widely condemned, and types of messages that may be widely accepted on one subject matter or from one government source may be criticized when the topic or source changes. A public official should be aware of the political risks and of the legitimate constraints placed upon government in advocacy and, where an unusually strong degree of advocacy seems warranted, seek political approval of such action.

Risk communication may be difficult because the purposes of messages are not clear or because they have multiple, perhaps conflicting, purposes. The next chapter describes several misconceptions about risk communication that may also contribute to confusion and frustration on the part of risk communicators and recipients.

NOTES

1. Generally, persuasive messages are less effective when recipients have the opportunity to "anchor" their preexisting beliefs against persuasion in the following ways: by defending them against a prior persuasive message, by considering their other beliefs or values that are supported by the belief subject to persuasive communication, or by training in the ability to question or argue against persuasive messages or to be suspicious of the source (the evidence is reviewed by McGuire, 1985:292–294). Persuasion that does not appear to be persuasion might not evoke such defenses.

2. For instance, public support for persuasive messages about AIDS prevention was minimal when the disease seemed to threaten only homosexual males, Haitians, and intravenous drug users but increased rapidly when children, hemophiliacs, adult heterosexuals, and hospital patients receiving blood transfusions were seen to be at risk. Shilts (1987) gives an extensive account of how public concern about AIDS has related to the identity of the groups believed to be at risk.

5
Common Misconceptions About Risk Communication

Some of the more important misconceptions about risk communication, including unrealistic expectations about what it can accomplish, are discussed in this chapter. Once these misconceptions are dispelled, the real problems of risk communication can be addressed.

We have taken care to distinguish between risk communication and risk management and between risk communication and risk messages. The primary goal of risk communication is to inform the participants in decisions about risks. Neither successful communication nor successful execution of the political process guarantees that risk management decisions will maximize welfare in terms of reducing exposure to hazards. Yet many people judge risk communication by the quality of the relevant risk management decisions.

We take political constraints as given and attempt to find ways within them to inform debates about risk. A well-informed decision process is likely to yield better decisions than an uninformed process. If all participants are adequately informed, the ultimate decision is more likely to improve conditions for all involved than a decision made by experts alone.

It is important, however, to realize that because risk communication usually involves multiple messages from many sources, and because these messages contain difficult and complex ideas, there is no simple way of making risk communication easy.

Risk messages necessarily compress technical information, which can lead to misunderstanding, confusion, and distrust. *Preparing risk messages can involve choosing between a message that is so extensive and complex that only experts can understand it and a message that is more easily understood by nonexperts but that is selective and thus subject to challenge as being inaccurate or manipulative.*

Since it is a reasonable precaution to assume that the compression in risk messages may introduce intentional or unintentional bias, it is natural to treat risk messages as reflecting political as well as scientific elements. Because people view risk messages as incorporating both scientific and political elements, appeals to scientific quality and veracity alone on the part of the risk communicator may not always sway the skeptic.

EXPECTATIONS REGARDING RISK COMMUNICATION

Many people—including some scientists, decision makers, and members of the public—have unrealistic expectations about what can be accomplished by risk communication. It is mistaken to expect improved risk communication to always reduce conflict and smooth risk management. In addition, risk comparisons alone cannot establish levels of acceptable risk or ensure systematic minimization of risk, although they can help people comprehend unfamiliar magnitudes.

Communication, Conflict, and Management

Many people, especially decision makers, seem to think that well-crafted messages or communication campaigns can eliminate or reduce conflicts in risk issues. These individuals believe that the conflicts are based on lack of information, that if all the parties were made aware of the facts, they would agree. This overlooks the possibility that conflicts are based on factors such as distribution of risks and benefits (e.g., do both fall equally on the same people?), different values (e.g., are the participants risk averse as opposed to risk seeking?), and different goals (e.g., is it better to avoid food additives or to enhance preservation and length of storage for food stuffs?).

Communication may reduce conflict about risks in some instances. However, when the underlying knowledge is uncertain, when people disagree about the meaning of existing data, when there is disagreement about the acceptable level of risk—in other words, in most cases of conflict about risk—informative risk messages might make the issues, and thus the conflict, clearer and more obvious.

In the Introduction we discussed the desire to develop effective alternatives to regulatory control as one of the reasons for interest in risk communication. But not all people see this as a positive development. The possibility of diverting attention from the risks and their control with careful information campaigns is sufficient to make some observers chary of risk communication. Ellen Silbergeld, senior scientist with the Environmental Defense Fund, expressed ambivalence about the large attendance (approximately 500) at the National Conference on Risk Communication in 1986. She viewed increased interest in the topic as a result of the destruction of consensus on environmental and other risk areas and described risk communication as a "shield for inaction" (Silbergeld, 1987a).

Comparing Risks

Another mistaken expectation is that risk comparisons can be used to determine acceptable levels of risk and help minimize overall exposures. Comparing different risks can help people comprehend the uncommon magnitudes involved and understand the level, or magnitude, of risk associated with a particular hazard. But comparison with other risks cannot itself establish the acceptability of the risk in question. To realize, for example, that the chance of death from a previously unknown risk is about the same as that from a known risk does not necessarily imply that the two risks are equally acceptable. Generally, comparing risks along a single dimension is not helpful when the risks are widely perceived as qualitatively different.

Risk messages commonly convey quantitative information that is unfamiliar and difficult to comprehend. These magnitudes and risk estimates are not easily understood without benchmarks or points of reference, and providing careful comparisons can help people understand this information. Risk magnitudes are difficult enough to understand when referring to a single consequence, such as death. But comparison of different consequences, such as injury, disability, or chronic disease, is even more difficult.

An interesting approach is the use of risk ladders, for which a range of probabilities is presented for a single class of risks. Although this technique can help people understand the magnitudes, it is not without problems. Figure 5.1 shows two examples of risk ladders. We consider the first weaker because of the several deficiencies listed. The second is considered stronger because it involves fewer deficiencies. The two risk ladders illustrate both the potential of the approach and the difficulty of using comparison. (Note: Not all attributes of the two risk ladders have been empirically tested, so it is not possible to state with certainty how people will react to them. The principal weaknesses listed are based on the existing literature. Each practical use of risk comparison should be carefully pretested if possible.)

Use of multiple comparisons helps counteract the possibility that people may severely misestimate a particular risk, even though it is familiar to them. It also reduces the danger of arousing the scientific disputes that can often arise when only two risk estimates are compared, one or both of which are subject to scientific debate.

One difficulty in risk comparison is that it is often difficult to find risks that are sufficiently similar to make the comparison meaningful. The easiest way to avoid comparing apples and oranges is to compare the risk associated with the same hazard at different times or risks associated with different options for achieving the same purpose. These comparisons are the least problematic because they address the same hazards and consequences with variation in the mechanisms for controlling or reducing the risk in question.

When such direct comparisons are not possible, it is important to recognize that various risks have different qualitative characteristics and that these can affect the way comparisons are viewed (Fischhoff et al., 1981a; Slovic, 1987; Slovic et al., 1980). Two that have been shown to have considerable impact are composite indices derived from factor analysis. The first, labeled "dread," is associated with perceived lack of control, dread, catastrophic potential, and fatal consequences. The second, called "unknown," is associated with the degree to which the risk is perceived to be unobserved, unknown, new, and with delayed manifestations of harm (Slovic, 1987). Hazards whose quantitative risks are estimated to be the same or similar may result in quite different responses if their qualitative characteristics are sufficiently different. Care must be taken that the risks compared exhibit qualitative characteristics that are reasonably similar.

Another pitfall of risk comparison is the appearance of selecting risks for comparison that minimize or otherwise trivialize the risk in question (Covello et al., 1988). Compendiums of risks, or risk ladders placing various risks along a spectrum from lower to higher, may give this appearance when the risk in question is much lower than other risks and when there are few risks presented with comparable levels. If, however, the comparison presents risks that clearly relate to the risk in question and relate or position its level or magnitude, the appearance of trivialization can probably be avoided.

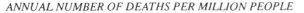

ANNUAL NUMBER OF DEATHS PER MILLION PEOPLE

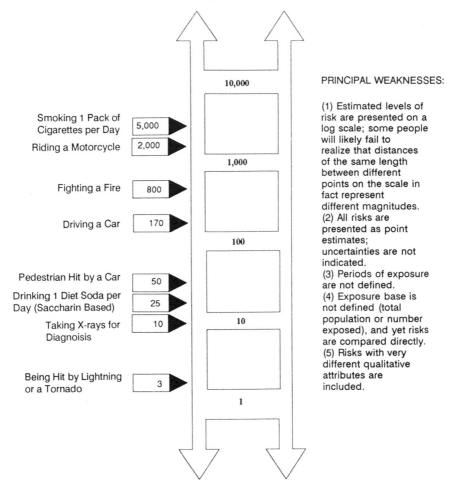

FIGURE 5.1a A poor risk comparison. SOURCE: Schultz et al., 1986, as cited in Covello et al., 1988. Reprinted with permission of the Chemical Manufacturers Association.

It is sometimes assumed that once they are told about risks people will systematically minimize their exposures and disregard truly small risks when they understand how little they are. This encourages comparing the risk in question to other risks that are familiar to most people with the intent of claiming that the level of the risk under examination is acceptable. The logic of using risk comparison to determine acceptable risk usually runs as follows: since you

RADON RISK CHART

Lifetime exposure (picocuries per liter)	Lifetime risk of dying from radon* (out of 1,000)	Comparable risks of fatal lung cancer (lifetime or entire working life)
75	214-554	
40	120-380	
20	60-210	Working with Asbestos
10	30-120	
		Smoking 1 Pack Cigarettes/Day
4	13-50	
2	7-30	Having 200 Chest X-rays per Year
1	3-13	
0.2	1-3	

PRINCIPAL WEAKNESSES:

(1) Exposure not defined (e.g., amount of time spent indoors).
(2) It is not known whether most people accurately perceived the anchors (i.e., asbestos, smoking, X-rays).
(3) The original uses colors that may be misleading.
(4) Does not use a linear scale.
(5) Uses anchor risks (i.e., asbestos, smoking, X-rays) with different qualitative attributes.

FIGURE 5.1b A better risk comparison. SOURCE: Smith et al., 1987, as cited in Covello et al., 1988. Reprinted with permission of the Chemical Manufacturers Association.

accept the risk of driving an automobile, which is about 240 annual fatalities per million persons (total population), you also ought to accept the risk of exposure to X (whatever hazard the communicator supports), which is, say, 10 annual fatalities per million. This logic is faulty (Fischhoff et al., 1981a). A homeowner, for example, should not neglect the potential fire hazard of electrical appliances or gas stoves and furnaces just because the risk of annual fatality due to fire is about one tenth as large as that due to driving an automobile. Rather, reasonable precautions should be considered with regard to risks deriving from all the hazards over which one has control. The level of risk is only one among several factors that determine acceptability (Fischhoff et al., 1981a; Gould et al., 1988; Slovic, 1987; Slovic et al., 1980), and the information requirements for an informed decision by private individuals or public officials will generally include more than the level of risk alone.

BELIEFS ABOUT THE FUNCTIONING OF THE PROCESS

Many problems for risk communication derive from mistaken beliefs about the nature of the risk assessment, risk management, and risk communication processes.[1] It is mistaken to expect scientific information to resolve all important risk issues. In addition, even when valid scientific data are available, experts are unlikely to agree completely about the meaning of the data for risk management decisions. Finally, it is unrealistic to expect easy identification and understanding of the values, preferences, and information needs of the intended recipients of risk messages.

Adequacy of the Scientific Information Base

As is clear from the discussion in Chapter 2, it is unrealistic to expect complete information about all the various aspects of a hazard and the risk of exposure to it. But even if the scientific risk information were perfect, it might not resolve all the issues involved. The best technical analysis cannot reveal what ought to be done. Analysis can only estimate the consequences and, in some situations, the way those expected outcomes compare to other related outcomes.

The adequacy of the information base is an important consideration not only because some statutes as well as current interpretation of the Administrative Procedures Act require regulatory decisions to be based on reasoned consideration of the evidence, but also because risk management decisions should be based on the best available information rather than arbitrary or unfounded beliefs and assumptions. It could thus be argued that the information base for a risk management decision would be inadequate if additional scientific data could provide at reasonable cost a more detailed or more complete understanding of the phenomena giving rise to the risk in question. Of course, more scientific data always would be of positive value under such a criterion, and the difficulty lies in determining how many resources should be allocated to this particular problem and how long the decision should be delayed in order to obtain more data.

Agreement as to the Meaning of Existing Information

There is seldom definitive scientific data about important risk issues. Science continually develops new, more sophisticated testing methodologies. Even with the most recent additions it is doubtful that any substance or product has been, or can be, so thoroughly tested as to preclude further scientific question. The numbers usually can only give an estimation of the consequences and, in some situations, the way those expected outcomes compare to other related outcomes. Very often regulatory and other risk control decisions must be taken before all the scientific questions are fully resolved. In these cases the decision maker will be faced with choosing from among conflicting, sometimes contradictory interpretations of the data.

These issues are important because they can strongly affect the determination of risk concerning a particular substance or activity. Whether a linear or multistage model is used for extrapolation, or whether a restricted or generalized model is used to compute doses, estimation of the no observed effect level (NOEL), or the safety factors used to allow for various kinds of uncertainty, can have significant impact on the characterization of risk. Such issues can be at the center of a controversy and can dominate debate about them and the related risk messages.

Interpretation of Public Attitudes and Information Needs

Because of the public's ability to make itself heard on risk issues, public opinion does influence the introduction and application of modern technology. But it is usually a relatively small part of the general public that makes its views known about a particular issue. It is therefore useful to distinguish between the passive public (largely unaware of the issue), the attentive public (aware of the issue and its ramifications), and the active public (seeking to make its views known or to affect decisions in other more direct ways). Depending on the nature of the issue, the source of a risk message may need to understand the attitudes and information needs of each of these different types of potential recipients of risk messages. Both the differences among these types of potential recipients and the respective ease or difficulty of establishing contact with them and determining their views and information needs contribute to the complexity of the task.

A few years ago several of the large groups of government and industry risk managers began seeking out the advice of social scientists because of opposition to their programs in the public (Fischhoff, 1985a). Risk managers typically made confident statements about public opinion on the basis of anecdotal observation, in contrast to practicing social scientists, who usually venture carefully qualified statements only after extensive investigation. Risk managers also made confident statements about the information the public wants and uses in particular situations. For the most part both types of statements were based on a view of "the public" that did not differentiate among the general public, the attentive public, and the active public or among people with different personal values, levels of exposure, or sensitivities to the hazards in question.

Not only does the level of interest in specific topics vary among different people, but so also does the way they think about the issues involved. During the last decade researchers have examined the opinions people express when asked, in a variety of ways, to evaluate hazardous activities, substances, and technologies (Slovic, 1987). Psychological research suggests that people's perceptions and attitudes are not determined solely by the sort of unidimensional statistics used to describe the magnitude of risks. To many people, statements such as, "the annual risk from living near a nuclear power plant is equivalent to the risk of riding an extra 3 miles in an automobile," give inadequate consideration to important differences in the nature of the risks from these two technologies. As noted in Chapter 3, risk is only one facet of these conflicts (see also Douglas and Wildavsky, 1982; Short, 1984). Risk concerns may provide a rationale for actions taken on other grounds or they may be a surrogate for other social or ideological concerns. When this is the case, communication about risk is off the mark.

STEREOTYPES ABOUT INTERMEDIARIES
AND RECIPIENTS

Some risk communication problems derive from misconceptions about the way intermediaries and recipients react to risk messages. It is mistaken to view journalists and the media always as significant, independent causes of problems in risk communication. It is mistaken, as well, to expect people only to want simple, cut-and-dried answers in every case.

Journalists and the News Media

Many people who are disgruntled with the slowness and apparent incoherence of decision making about risk control or with the outcomes attribute many of these problems to the news media and journalists. Some claim, for example, that public concern is "driven by media coverage rather than by rational scientific analysis" or that "the media has driven the public insane" (Cohen, 1987). These critics claim, for example, that the news media are basically in the entertainment business and that the only thing that matters is the ability of a story to attract attention because this sells newspapers and attracts viewers.

It is true that newspapers, radio and television stations, and the networks are businesses. And it is true that they must pay attention to income and profits. But the direct effect on subscriptions or advertising income is not likely to be in the minds of reporters as they prepare stories nor in the minds of editors or producers as they make story assignments, edit copy, or determine the placement of various stories in that day's newspaper or newscast.

In selecting and preparing stories, the reporter is much more likely to be motivated by events, by what other reporters are paying attention to, by information provided on a regular basis by sources he or she has cultivated, by deadlines, and by what interests him or her as a citizen. The editor or producer will be concerned about the appeal and impact of the issue or program as a whole. Both, for their different reasons, will be concerned about the importance of the stories, their impact, and their drama. The attractiveness of stories with such appeal will be strong whenever censorship is absent and there is free and open access to information sources. It is mistaken to attribute the way the media defines "newsworthiness" in practice to crass economic motives alone.

Because of their involvement in selecting and preparing stories, journalists may have a better perception of the audience and its interests than do editors or producers. But that perception is probably also based on the "convenience sample" with which that journalist happens to have contact. Journalists and the media play important roles in revealing conflicts and sometimes in their resolution. The media may nurture the development of controversy by serving as a channel for debate among the major actors in a conflict, and they can play crucial roles in providing information to citizens during conflicts (Tichenor et al., 1980). This latter can be especially important since significant portions of the public may never attend to risk information unless such a conflict attracts their attention.

For the most part, what can be called the national press (the *New York Times, Wall Street Journal,* and comparable news organizations) treats risk issues with considerable care and understanding. But this is not always true, especially at the regional or local level. The performance of the press in the reporting of risk issues is not always up to the standards found in other topic areas. Most news organizations would not tolerate sports or business reporting by reporters who do not understand the subject and are unable to correctly frame those topics. The same is not always true of the reporting of the technical and social dimensions of risk messages.

Some criticism of the news media emerges from a failure to examine the structure of the media industry or how journalists work. It would be more fruitful for risk communicators to try to understand the pressures and constraints on news gathering than to curse the sometimes disappointing results. The structure of the industry and the incentives and influences that affect the way it works are part of our social and political system. What is needed are ways to improve risk communication by helping scientists and decision makers understand how and why journalists do their work and by helping journalists understand how scientists and decision makers think and interact.

There are, for example, differences between the structure and incentives affecting the broadcast media and those affecting the print media. Material with visual impact will be especially appealing for television. There also will be differences within segments of the different media. The focus and approach of science magazines, for example, differ from those of straight news magazines. National newspapers differ from regional or local newspapers. Despite these differences, however, the overall impact of the incentives and influences on reporters, editors, producers, and so on is more similar in the various media than different.

Another characteristic of the press worth understanding is that most reporters deal with news, not education (Sandman, 1986). It is usually the events that make something newsworthy, not the issues or principles involved. News reporters seldom want to know the ins-and-outs of risk assessment, how sure the experts are, or how they found out. They want to know about the number of people affected; the gravity of the consequences; and the cost of damage, repairs, or remedies. They pay attention to the vividness with which these can be presented.

Feature stories, such as those found in Sunday editions or special broadcasts, can go into much greater depth and offer more complex treatments. Specialist reporters also pay attention to newsworthiness, imagery, and so forth, but they tend to go into greater depth in laying out the background and some of the underlying factors that bear upon the events.

Most journalists care about accuracy and objectivity. Often the only operational definition of objectivity for journalists is balance (Sandman, 1986). They are seldom experts in the topics they cover. They cannot, as a result, determine for themselves what is true. They can only try to present the conflicting claims fairly. And because their job generally is reporting events rather than issues, they get most of their information from people who are directly involved in the event and only occasionally seek out uninvolved experts for advice. Some journalists, especially at the regional and local levels, also emphasize the reactions of ordinary people. They present the events of concern, the consequences and their importance, any conflict about outcomes or responsibilities, and the response of "the man on the street." This helps people interpret the news in terms of themselves, their families, and their neighbors.

Journalists may seek out those with conflicting claims about the events in the news. In striving for a balanced coverage, they often attempt to identify extreme positions about the events or issues. Not being able to assess which positions have been given greater credence among the community of experts, they attempt to discover the range of views. Although they may not present the most extreme positions—the ones and sevens on a range from one to seven—they will typically look for individuals expressing well-defined positions that bracket the middle of the range of relevant views—the twos and threes and the fives and sixes. Positions that clearly differ in this way are attractive to the journalist because they define the range and because their juxtaposition sharpens the drama and heightens interest.

To be sure, there have been instances in which media coverage has favored one extreme, such as the television network that showed a skull and crossed bones in the background whenever a reporter spoke about ethylene dibromide (EDB) (Sharlin, 1987). But there is also some evidence that even in events with massive attention and media coverage, the news media seek balance. An extensive content analysis of media coverage of the nuclear industry accident at Three Mile Island found the balance between supportive and negative statements to be, if anything, more reassuring than alarming (*Report of the Public's Right to Information Task Force*, 1979; Stephens and Edison, 1982).

The Attraction of Decisive Answers

The public often appears to want decisive, clear-cut determinations of risk and descriptions of the appropriate control measures, especially when the choices they face appear to be simple dichotomies—a product can be used or not used, an incinerator built or not built. This response is based on fundamental psychological mechanisms. Most people prefer simplicity to complexity in matters outside their own field of expertise. In addition, most people are too busy to spend much time on any particular topic, and some find it hard to understand why information about risk cannot be put in concise, decisive terms. Unfortunately, one seldom knows how often or in what mixes these various situations obtain.

Sometimes, however, people prefer to have the options laid out for them and to be given the choice of selecting the one they prefer. This is most common when the risk control measures require action by the individual. Examples include using seat belts, choosing among medical treatments, and changing sexual practices to curb the spread of contagious diseases such as hepatitis or AIDS.

Several things may influence people's preference for decisive or ambiguous information: the degree to which they as individuals exercise control over exposure or remediation, the importance they attach to the issue, and their tendency to be risk averse or risk seeking. That different segments of the population may prefer decisive or equivocal information about a particular risk can make the job of the risk communicator more difficult. It may even be that individuals prefer different types of information at different times during the course of discovery, analysis, and control of a hazard.

This chapter has discussed some of the more important misconceptions about risk communication. The next chapter addresses directly what we believe to be the most important problems confronting the practice of risk communication.

NOTE

1. These and other relevant terms are defined in a list given in Appendix E.

6
Problems of Risk Communication

In this chapter we address what we consider to be the principal problems of risk communication. First we describe problems deriving from the structure of the political and administrative system. These are problems for which little can be done by those involved in risk communication beyond understanding them. They must be confronted and accommodated, since they cannot be done away with.

Next we describe problems of risk communicators and recipients. These problems, in contrast, are much more amenable to improvement or solution. The problems of these two groups are presented together because many things are problems for risk communicators because they are problems for the recipients of risk messages. For example, the risk communicator needs to pay attention to the understandability of risk messages because most recipients have difficulty comprehending the technical terms typically found in risk assessments and other technical analyses.

PROBLEMS DERIVING FROM THE INSTITUTIONAL AND POLITICAL SYSTEM

As we have seen, scientific and technical information is of central importance to decisions about how to respond to risks and thus is an important element in risk messages. But risk management decisions also take place as part of a democratic process, and risk analysis is only one of several sources of relevant information. Furthermore, politics can, and often does, assert control over decisions otherwise delegated to experts. The intrusion of politics can result in considerable frustration to risk managers as well as to others involved in the process. It is thus important to consider problems posed by the institutional and political system for risk communication.

For the most part the problems of the institutional and political system are part of the context within which risk managers and risk communicators operate. Even though these problems are largely beyond the ability of the principals in risk communication to affect, they nevertheless can have considerable impact on actions and events.

Legal Considerations

Risk communicators may be constrained because legal considerations influence the options available to risk managers and therefore the content of risk messages. Several kinds of legal provisions may provide such constraints, including (1) statutory mandates, (2) liability, and (3) informed consent and "right-to-know" requirements.

Statutory Prescriptions and Proscriptions

Statutory language may, in effect, force the risk manager to take certain kinds of actions, some of which have important consequences for the content of risk messages or their dissemination. This is perhaps most obvious with respect to units of the Public Health Service. The major goal of the Centers for Disease Control (CDC), for example, is to lead public health efforts to prevent unnecessary disease, disability, and death. The CDC pursues this goal through programs aimed at prevention and control of infectious and chronic disease and of disease, disability, and health associated with environmental and workplace hazards (Department of Health and Human Services, 1986). These programs include not only regular publications such as the *Morbidity and Mortality Weekly Report* (MMWR) but also emergency advisories. CDC officials are often quoted in the news, and their statements can have considerable impact. CDC's concern about long-term contact with soil in the Times Beach, Missouri, area was an important factor in the government's decision to purchase homes and permanently relocate the residents. CDC's mandate to lead public health efforts thus goes a long way toward establishing the tone and approach of risk messages emanating from the CDC.

Such prescriptions and constraints and their impact on action may in their turn have a strong impact on whether the public views the agency as an advocate and on the credibility of the organization with respect to public health issues.

Liability

One principal reason that legal constraints constitute a problem for risk communication is that these considerations may make difficult or impossible the crafting and presenting of messages that effectively address the issues that may be most relevant to the intended recipients of the message. For example, following the 1985 release of aldicarb oxime from its plant in Institute, West Virginia, Union Carbide had to decide about what information to make public about the accident (Coppock, 1987). Communications and community relations experts usually advise making available everything that is known about an accident as quickly as possible, in terms that laypeople can readily understand. Legal advice is almost always exactly the opposite: give out as little information as possible so as to avoid providing ammunition for use in court. Given the prevalence of large court awards in product liability and toxic exposure cases, concern with liability is in the minds of many business people. The final message probably involves a compromise between these perspectives.

Informed Consent and Right-to-Know

Issues of informed consent have changed the way the health profession interacts with patients. Attempts are made, for example, to hold physicians to fairly stringent standards in obtaining consent prior to initiating experimental therapy. But "right-to-know" issues are having equally important impact in many other areas. Employees are to be informed about the hazards of the materials they handle under Occupational Safety and Health Administration rules, communities are to be informed about inventories and emissions of hazardous substances under the community right-to-know provisions in Title III of the Superfund Amendments and Reauthorization Act of 1986, and California's Proposition 65 provides for provision of information about any product containing carcinogens or teratogens. The overall effect of such developments is that there are many more legal requirements that result in the preparation and dissemination of risk messages than in the past.

Sharing of Power

Communicating with citizens about risk issues can increase their desire to participate in or otherwise influence decisions about the control of those risks. These demands can change the dynamics of the situation for the risk manager, the risk communicator, and the citizen. The motivation for citizen involvement becomes even stronger when a decision process appears to result in an outcome with which the individual disagrees. The interests of citizens and their motivation to participate can be especially problematic when the implementation of risk control measures is necessarily decentralized and local preferences preclude solutions in the broader interest.

The sharing of power is a central facet of representative democracy. Citizens transfer decision-making power to elected officials. However, citizens have the responsibility to hold both the legislative and the executive branches of government accountable. To exercise this responsibility and judge the delegation of authority, citizens need information. Demonstration of this accountability is one of the important functions of risk communication.

Holding government accountable means, in a basic sense, ensuring that government policies and actions correspond to public preferences. The difficulty, of course, is aggregating across the preferences of the many people involved. Most people believe they have a right not to be subjected by others to unreasonable risks. Some people believe not only that risk to life and health should be minimized but also that three kinds of unfairness should disqualify, say, siting a hazardous waste facility: (1) imposing costs on those who have not voluntarily agreed to bear them, (2) imposing costs on those who oppose availability of and avoid use of the products and services generating the hazard, and (3) imposing disproportionately large burdens on those who benefit least (Simmons, 1984). When people believe that any of these three hold, they may feel imposition of a hazardous waste site to be unfair regardless of the processes used to derive that particular site. Such conflicts are at the core of many instances of "locally unwanted land uses" (LULUs). Local preferences often run counter to solutions that would otherwise seem to be in the broader public interest. Here risk is only one part of the problem, and thus risk communication per se cannot be expected to resolve all the issues.

Communication research suggests that risk messages will be more easily understood when the risk communicator not only incorporates language familiar to the recipients but also genuinely respects and incorporates their views (Covello et al., 1987b). This perspective suggests that effective risk communication should begin before important decisions have been made. If all the important choices have already been determined, it will be difficult to reflect the views of the recipients. Risk messages will become little more than attempts to "sell" a predetermined conclusion, which may create considerable alienation among the intended recipients. But as suggested in the above discussion, people may reject the attempt no matter when it occurs if they are unwilling to compromise their position. The risk manager may thus face a difficult task in seeking advice from people but excluding them from the resulting decision.

People naturally want to see their views affect the outcome, and they may have difficulty differentiating the risk communication process from the risk management process. The American political culture puts a premium on procedures that offer a wide variety of interest groups and citizens the opportunity to participate in decision making (Melnick, 1988).

Fragmentation

Risk control decisions can be made or influenced by several different political actors. At the federal level, Congress, the executive branch, and the courts all shape health and environmental regulations. Despite the dominant role of the federal government, state and local governments also remain important. This fragmentation may make communicating about risks more difficult because of dispersion of responsibility, incentives for each actor to gain as much leverage as possible from the limited portion he or she controls, and difficulty in determining who is responsible for the eventual outcomes.

Dispersion of Responsibility

Fragmentation of risk control decisions derives from a central feature of the structure of American political institutions: dispersion of power. A basic tenet of the American political system is separation of powers, but power is also dispersed to a remarkable degree. For example, one source claims that Edmund Muskie, though only chairman of a Senate subcommittee, had at least as much influence on environmental policy from 1969 to 1979 as Richard Nixon, Jimmy Carter, William Ruckelshaus, or Douglas Costle (Melnick, 1988). And several environmental groups, most notably the Natural Resources Defense Council and the Environmental Defense Fund, have used their success in litigation to become major players in national policymaking (Melnick, 1988).

In the United States each level and branch of government provides access to a variety of groups. Thus corporations, trade associations, labor unions, professional associations, intergovernmental lobbies, and environmental groups all influence regulation and its implementation. To the extent that individuals and organizations participate, they can also be held responsible for the overall outcome.

Dispersion of responsibility can lead various executive agencies to take different positions with respect to the same issue. The U.S. Department of Energy, for example, views the hazards associated with radioactive contamination of groundwater differently than does the EPA. Officials of the two agencies say quite different things about the risks involved in specific instances of contamination in Idaho. Differing positions can also be found within different parts of the same government organization. This derives in part from the organization of large bureaucracies into separate divisions but also in part from the belief that separation of power yields greater benefit than cost.

When fragmentation leads various parts of government to different positions or approaches with respect to the same risk, it can lead to problems for risk communicators.

Incentives to Gain Leverage

The extensive dispersion of responsibility among parts of government means that there are often jurisdictional conflicts and overlapping responsibilities among different governmental organizations. The existence of these overlaps can provide the opportunity for particular organizations to apply leverage beyond their organizational boundaries. The "crisis" involving groundwater and contamination of foodstuff with ethylene dibromide (EDB) is one example. At the federal level the EPA, the Food and Drug Administration, the Department of Agriculture, and the Occupational Safety and Health Administration all had responsibility for some aspect of exposure to EDB. They had more or less reached agreement as to the handling of the pesticide. But action by state government agencies in Florida and Massachusetts brought EDB to the public attention and forced changes in the response from the federal government. The Massachusetts EDB team leader, Dr. Havas, summed up the state

Department of Public Health experience like this: "It was a success
. . . particularly how quickly we got EDB out of the Massachusetts
food supply. What we did drove EDB out of the food supply for the
entire nation . . . not just for Massachusetts . . . everybody got the
benefit" (quoted in Krimsky and Plough, 1988).

Difficulty in Determining Responsibility for Outcomes

Dispersion of responsibility and the actions of various individuals and organizations to obtain as much leverage as possible can
mean that the recipient of risk messages has a difficult time knowing exactly which organizations have jurisdiction over the hazard in
question. Various organizations may have competing aims or goals
with respect to the hazard in question and the control of the associated risks. The resulting confusion can constitute a problem for the
risk communicator because he or she needs to clarify the organizational responsibilities as well as the risk involved. The fragmentation
of risk control decision making thus contributes to the difficulty of
communicating about risks.

Imbalanced Access to Information

*If the group of people that a risk communicator is trying to reach
thinks that the system for generating information relied upon by that
source does not consider its concerns, it may reject the information
from that source as a basis for decisions about risks.* Rejection of
its information can be a considerable problem for a risk communicator. Organizations disseminating risk messages need to be aware
of the effects of uneven access to information by those affected by or
requesting the organization's action.

Information is not free. It is expensive to develop empirical data,
and there are not enough research funds to examine all questions
that might be relevant to particular issues. Thus the amounts of
information about all considerations relevant to such decisions are
unequal and may therefore introduce imbalance into the information
base for risk decisions.

Government and industry spend large amounts of money on research. This not only encourages their concerns to be reflected in
research projects but also establishes patterns of information flow
and interactions that reinforce this effect. Environmental groups or
trade unions do not have equal amounts of money to fund research

and may be at a disadvantage in justifying their positions in conflicts about regulatory decisions or other risk management strategies. However, they can often serve the valuable function of criticizing the information developed by other organizations.

Science tends to be conducted in institutional settings with strong incentives—the amateur scientist was, for the most part, a character of the last century. Researchers at universities and other independent research facilities are subject to powerful influences, both through budgetary constraints and the need to publish their results in peer-reviewed journals. There never will be enough research funds to pursue all questions relevant to particular hazards. Funds that do exist may be inappropriately allocated. Issues that are popular in particular disciplines may thus introduce imbalance into the information base for risk decisions.

Even when information has been created, it may not be equally accessible to everyone. The research community can be reached more easily by those with resources to support, follow, and interpret its activities.

Local citizens' groups are likely to have even less contact with relevant research communities. They will probably be unfamiliar with the language of science and may not formulate their questions in ways that scientists can use. This may detract from the usefulness of public hearings and other settings where exchange might take place between the providers of information and concerned citizens.

Systematic Interests and Biases

Those most strongly motivated to communicate about risk are often also those with the strongest interest in the decisions. So whenever a personal or a social decision may affect interested groups, conflicting messages that reflect the conflicting interests may be expected. The beliefs of risk communicators, and their interests, create incentives to slant or even distort or misrepresent information. This can skew messages in many different directions on the same issue.

The American Cancer Society and the Tobacco Institute offer conflicting messages about the health effects of smoking, the National Agricultural Chemicals Association and the National Farmworkers Union are in conflict about the health risks of pesticides, and the Sierra Club and the Edison Electric Institute take different positions about the dangers of acid rain. The reasons for these differences may be complicated, but smokers contemplating quitting, farmers considering the adoption of integrated pest management, and citizens taking positions on the regulation of air pollution are confronted with making judgments about the risks by weighing messages from obviously interested sources and messages from other sources whose biases are not so obvious.

Consider the experience with messages about AIDS. Fearing the response to the epidemic of a traditionally homophobic society, various groups representing the gay community have at different times underplayed and exaggerated the risk of AIDS (Shilts, 1987). Initially, the gay community denied that there were special risks associated with homosexual practices and sought to protect bathhouses and other gathering places from interference by public health officials. As the toll has increased, the tendency has been to claim rapid spread of the disease among heterosexuals. Gay community groups tended to describe AIDS as a societal affliction not concentrated in an isolatable and stigmatized group. When everyone is a potential victim, both compassion and resources are likely to become more plentiful.

For their part, blood banking organizations have consistently sought to underplay the risks of AIDS contracted through transfusions. A prime motivating factor has been their need to maintain an adequate supply of blood for the nation. If blood is linked to a new and highly dangerous disease, the public might, as has happened, curtail donations in a mistaken belief that there is a risk to donors. Until 1982 the blood banking community rejected epidemiological evidence that AIDS could be transmitted through banked blood and told the public the blood supply was "safe," when all that was known was that the risk of AIDS had not been convincingly demonstrated. The overriding concern was a desire to reassure the donating public (Holland, 1987). The blood banking community continues to claim that the blood supply is "as safe as it possibly can be for AIDS," although some recommend that additional screening procedures be used (Holland, 1987).

The point is that on matters of public controversy risk messages tend to be flavored by the positions taken by the sources of the various messages. Moreover, these biases are not necessarily obvious to those who receive the messages and use them to make personal decisions or to inform their political positions.

PROBLEMS OF RISK COMMUNICATORS
AND RECIPIENTS

Examinations of risk communication have tended to focus on the preparation, presentation, and transmission of messages about the nature of risks and risk reduction measures and on their receipt and interpretation by the intended recipients (Covello et al., 1987b; Davies et al., 1987). Most of this attention has been directed at the problems of the individual or office preparing and disseminating risk messages. Here we describe many important problems the risk communicator will face in these tasks as well as the special problems of the recipients of such messages. We also examine aspects of the interactions between the risk communicator and other groups: other people within his or her organization, other groups or organizations, and the intended recipients.

One of the central aspects of risk communication is that risk messages are not created and transmitted in a vacuum. The policy, administrative, or political arena within which the communication process occurs is an important influence on what eventually happens. We describe problems that derive from within the risk communicator's organization or group as well as those that characterize the broader setting of interactions with other individuals, groups, and organizations.

Debates between risk managers and experts, or between experts and members of the informed and involved public, are often poorly understood by the general public. Although such debates are not particularly well attended to, they are also not ignored. Risk debates often are interpreted by the general public in two ways: the world is a dangerous place, and risk managers either do not know what they are doing or do not understand what they are supposed to be doing. In other words, risk debates often generate fear, which is unpleasant and generally not helpful for making decisions. Neither heightening of public fear nor heightening of public distrust of risk management can be considered constructive as such. But even though risk communication may engender at least some fears in the public regardless of content and procedural safeguards, we feel that it is a necessary and important part of risk management in a democracy.

The risk communicator attempts to present information in such a way that the intended recipients will receive and attend to its message. Usually, the risk communicator presents this information in the hope of influencing the recipients' attitudes or actions. But the recipient may not particularly care about the issues raised by a message. Many messages are likely clamoring for his or her attention, and those about the same issues are likely to be contradictory. The recipient is faced with the difficult task of making sense out of a confusing mess of information from many different sources.

To some extent, the problems of the recipient of a risk message are the mirror image of those of the risk communicator. This is the reason we address them together. The communicator worries about credibility because the recipient judges messages on the basis of the reputation of the source as well as the content of the message. Risk messages not only need to be, but must also appear to be, accurate and responsible representations of the issues because the skeptical recipient will be on the lookout for incompetence, inaccuracy, misrepresentation, and deceit. Similarly, the communicator tries to be clear and easily understandable because most recipients have difficulty with complex technical material.

Occasionally recipients of risk messages become risk communicators. When an individual becomes motivated to join or create a group with the aim of influencing decisions about risks, he or she generally disseminates oral or written messages to others. In these situations that individual will experience not only the problems of interpreting risk messages from other sources but also many of the problems of risk communicators. The risk communication process then becomes interactive in its most fundamental sense.

In examining the problems of risk communicators and risk recipients, we describe several things that make easily understandable risk messages difficult to achieve. We present general conclusions about mistakes to be avoided. The attributes of risk communicators and risk messages we identify and their impact on the risk communication process are derived from this general research base and our collective judgment. The research is much weaker, however, in giving guidance about what will work in specific situations. The only way to be sure is to pretest communications materials with representatives of the intended audience.

Establishing and Recognizing Credibility

Lack of credibility alters the communication process by adding distrust and acrimony. The most important factors detracting from the credibility of a risk message relate to the accuracy of the message and the legitimacy of the process by which its contents were determined, as perceived by the recipients.

The perceived accuracy of a message is hampered by the follow-
ing: real or perceived advocacy of a position not consistent with a
careful assessment of the facts; reputation for deceit, misrepresenta-
tion, or coercion; previous statements or positions that do not support
the current message; self-serving framing of messages; contradictory
messages from other sources; and actual or perceived professional
incompetence and impropriety.

The perceived legitimacy of the process by which the contents of
the message were determined depends on the following: legal standing,
justification of the communication program, access of affected parties
to the decision-making process, and fair review of conflicting claims.

Real or Perceived Advocacy of Unjustified Positions

Perhaps the most critical element of credibility for a source is
the degree to which intermediaries and the ultimate recipients of
the risk message believe that source to be justified in the position
reflected in the message. As we have already pointed out, it is
extremely unlikely that the recipients will be in the position to judge
the accuracy, balance, and fairness of a risk message from the content
of that message alone. One result is that recipients tend to judge
the messenger as well as the message. The reputation of the source,
in terms of past record with regard to accuracy of content and
legitimacy of the processes by which it is developed, will be an
important influence on the way recipients view particular messages.

It is important that an organization ensure that its positions
are technically competent. An unfortunate example of the failure to
do so involves the EPA and its decision to conduct a chromosome
damage study of the residents of Love Canal in New York (Levine,
1982). At the outset the decision was made, apparently by EPA
lawyers, to restrict the number of people studied because of the high
cost of studying chromosome damage (Davis, 1987). Individuals were
therefore selected to maximize the likelihood of damage, following the
reasoning that comparison of what should be a high-risk group to a
group from an uncontaminated neighborhood should make it easier
to determine whether a larger study would be justified (Levine,
1982). However, funds were further curtailed, so that a total of
only 36 cases could be included and the control group from the
uncontaminated neighborhood was eliminated. This ensured that
the results would be difficult to interpret, which was recognized
by the scientist conducting the study. Given the extreme emotions

surrounding the events at Love Canal, the result was considerable controversy. Five independent reviews of the chromosome study were submitted to the EPA, two requested by federal agencies and three by the scientist conducting the study. All emphasized the limited inferences that could be drawn due to the lack of a control group. The reviews commissioned by federal agencies criticized the interpretation of the data on chromosome damage in the study, while those requested by the scientist conducting the study were more favorable concerning the data interpretation. Although this example is extreme, scientific studies are subject to strict examination of their methods of data collection and interpretation. This examination is usually severe when the studies are used to support controversial public policy decisions.

Reputation for Deceit, Misrepresentation, or Coercion

Perhaps the most difficult problem for credibility is a past record of deceit, misrepresentation, or coercion. For example, as was acknowledged to us by officials from the U.S. Department of Energy (DOE), one of the biggest problems confronting the civilian radioactive waste program at DOE is the legacy of the Atomic Energy Commission and even earlier government programs (Isaacs, 1987). The attribution during the 1950s of fallout in St. Louis to Russian sources when in fact it was known to come from tests in Nevada was a blatant abuse of public trust, the repercussions of which the DOE must live with today. When the responsible government organizations have been proven to lie, it is not surprising that people want independent verification. One year of being honest with the people is not enough. Given the knowledge today of the cavalier treatment of facts concerning its activities in the past and the tremendous opportunity for uncertainty to enter its analyses and for its analyses to be skewed, the DOE faces tremendous credibility difficulties. Even the slightest indication of less than complete candor and honesty will probably lead many people to reject whatever position the agency takes. Given the highly politicized issues that DOE's program addresses, this legacy adds to an exceedingly difficult challenge.

The situation is somewhat different for nongovernment organizations. Private corporations, advocacy groups, and private citizens are commonly expected to interpret the facts of the situation in ways that support their aims and goals. This is part of the reason corporations and their messages are distrusted. Despite the difficulty that recipients of risk messages have in recognizing misleading or deceitful messages, a reputation for consistently bending the facts to fit one's purposes will undermine one's credibility to many recipients, although one's direct constituency may become more supportive.

Contradiction of Previous Positions

Establishing and defending credibility is difficult when the message represents a departure from previous positions. In large part credibility derives from the demonstration over time of consistent competence and fairness. Both scientific incertitude and changes in policy can serve to undermine credibility to the lay public. The necessity of correcting mistaken statements or positions can undermine credibility with the public. Care must be taken to demonstrate why the interpretation of scientific or policy conclusions has changed.

The rapidity with which new scientific findings about the AIDS virus that counter or revise previous positions are being presented undermines the credibility of the experts. At least one response to the rapid-fire changes in estimates and contradictory conclusions about heterosexual transmission of the HIV virus is to conclude that the experts really do not understand what is happening and that every imaginable precaution is thus justified.

Inconsistencies can also result from changes in administration, as was clearly illustrated by the treatment of formaldehyde by the Occupational Safety and Health Administration (OSHA). In December 1980, OSHA, acting in conjunction with the National Institute for Occupational Safety and Health, released a Current Intelligence Bulletin recommending that formaldehyde be considered a potential carcinogen and that appropriate controls be implemented to reduce employee exposure to the chemical. In March 1981 the new Assistant Secretary of Labor for Occupational Safety and Health rescinded OSHA's sponsorship of the bulletin. In January 1982, OSHA denied a petition by labor unions for an emergency temporary standard to reduce formaldehyde levels in the workplace (Ashford et al., 1983). Such reversals of position based on the same evidence can only reinforce the appearance of inconsistency and undermine the credibility of the source.

Self-Serving Framing of Messages

The leeway that exists in the collection of data, its interpretation, and the final crafting of messages provides ample opportunity to present information in ways that support the position of the organization more strongly than the evidence itself might justify. Such framing of messages need not involve direct deception or lying. For example, on August 27, 1986, the *Washington Post* reported on the Soviet government analysis of the potential health effects of the Chernobyl disaster under the heading, "Chernobyl Report Surprisingly Detailed but Avoids Painful Truths, Experts Say" (Smith, 1986). The "painful truth" was that the disaster may result in 35,000 to 45,000 cancer deaths in the Soviet Union and that "as many as 90,000 people could be affected by the recent explosion." But the Soviet report said that fatalities would be "less than 0.05 percent in relation to the death rate due to spontaneously arising cancer." Since this percentage works out to be 35,000 to 45,000 premature deaths over the lifetimes of the people exposed, both reports are equally accurate. But the two messages stimulate very different responses in most people.

The incentives to slant the presentation of information to support an issue one believes to be important can be strong. But in order to strengthen their credibility, public service organizations, and especially those in government, must resist this temptation.

Contradictory Messages from Other Sources

The adversarial nature of the American regulatory system is often cited as one of its strengths. But it also can help undermine the credibility of sources of risk messages. Parties with potential gain or loss are motivated to develop the best evidence and strongest arguments for their respective positions. When this is accomplished according to strict rules of evidence and scientific review, it can be expected to produce a reasonably complete picture of the issues in question. However, it also encourages competing interpretations of the evidence whenever there is uncertainty in the data, in applicable methods, or in the models for interpreting empirical results. Sometimes science is claimed to support all sides of a conflict about risk.

Conflicting messages also can derive from sources that are usually not associated with different "sides" in an adversarial situation. For example, state officials in Florida and Massachusetts sent clear signals that ethylene dibromide (EDB) should be of serious concern at exactly the same time that the EPA administrator was attempting to reassure the public (Krimsky and Plough, 1988).

In most situations the risk communicator will have to deal with conflicting messages from other sources. Contradictory messages can be a central part of the issue as seen by the public or may be a relatively minor side point. If the communicator does not deal with these effectively, it can undermine his or her credibility.

The extremes the recipient needs to watch out for are intentional manipulation and outright dishonesty. Although it is generally difficult for the layperson to determine from the message itself when it is manipulative or dishonest, some are so poorly crafted as to be blatantly miscreant. Less extreme instances are exceedingly difficult to identify. One strategy is to see how other individuals and organizations with a stake in the issue respond to the positions and statements of that source. The other parties in an issue have incentives to sort out misrepresentations and unsupported findings of their opponents and make them known.

The difficulty of determining the degree to which a risk message reflects advocacy of a particular position can be illustrated by reference to the experience with EDB in the early 1980s (Krimsky and Plough, 1988). Environmentalists critical of the basic federal regulatory treatment of pesticides found the debate on EDB to be an excellent opportunity to press their argument. They were joined by some state officials, who used the issue as a lever to force the federal agencies to act. Industry and trade associations thought the issue illustrated the need to weigh economic considerations in the regulation of hazardous chemicals. The EPA had to both defend its previous decisions with regard to EDB and quickly evaluate new data on exposure levels. The news media attempted to package all this in a way that would be newsworthy. And the recipient had to weigh all these positions against each other.

Another problem for the recipient of risk messages is to differentiate conflicts based on scientific disagreement about the facts or their interpretation from conflicts based on advocacy of policy aims. Sometimes this requires determination of the extent to which the facts reflect advocacy. If the source has identified its aims and purposes, this task can be made easier. Unfortunately, many sources are much more concerned with the outcome of the issue under question than with helping the recipient understand all its ins-and-outs. In fact, some sources deliberately confuse such questions and obfuscate rather than clarify because such actions contribute to the possibility that their position may hold sway. The recipient of risk messages may thus have considerable difficulty separating scientific conflicts from policy conflicts.

One of the effects of "dueling experts" may be to reduce the importance of expertise in the minds of the recipients of risk messages. If scientific evidence is inconclusive, then why not use whatever factors seem to suggest a solution? One result of such an approach would probably be to shift the nature of the problem definition away from the aspects of risk that have been quantified to other factors that may or may not be measurable. This has probably happened with respect to nuclear power. The technical questions are no doubt less critical to most people than their beliefs about the overall impact of increasing or decreasing reliance on nuclear reactors.

Professional Incompetence and Impropriety

A major element helping determine the credibility of risk messages is the perceived competence of the individual or organization concerning the subject under question. An individual with special training about the phenomena involved is often accorded greater credibility than someone whose training is less relevant. The statements of a physician, for example, might be given greater credence with respect to a public health question than those of a dentist or health economist, even though each is a health professional. Similarly, organizations enhance the credibility of their messages about technical issues when they have professional staff with training in the areas covered. Two strong criticisms of government agencies are that they do not have sufficient staff with the necessary professional competencies to fully and completely understand the phenomena they regulate and that the understanding that the professional staff does have is not shared by the nontechnical personnel at the apex of the organization.

Private organizations, especially business corporations, are more often suspected on account of propriety than competence. Many people believe that corporations have, or hire, the best expertise available but that these experts present only information that is in the interest of the corporation. Since the decisions of private organizations are not subject to the constraints of due process as are those of government, there is a strong presumption by many that the messages of private organizations emphasize only the information they believe to best serve their interests. Similar descriptions apply to many public interest or public advocacy groups.

There are thus strong incentives for an organization disseminating risk messages to be as open and clear as possible about the way it gathers and interprets information. This is especially true if it incorporates scientific peer review or other technical review procedures. Demonstrating the professional competence of the organization and the propriety of its procedures will likely enhance the way its messages are received.

Legal Standing

An important influence on perceived credibility is legal standing and involvement in the issue under question. Being lawful or corresponding to the dictates of the law is a fundamental component of legitimacy.

As we have indicated above, a special sense of responsibility attaches to government. Government officials and agencies are expected to act in the public interest. At the most fundamental level the legitimacy of a federal agency to speak to an issue derives from its statutory mandate and from the exercise of due process under the Administrative Procedures Act. Determining this is relatively straightforward, although there may be questions of jurisdiction among different agencies. For example, there may be disagreement about whether the EPA or the OSHA should regulate airborne toxic contaminants in particular situations.

On occasion, however, government agencies are expected to act and, if appropriate, interact with the public even when the topic is not strictly within their statutory responsibilities. For example, the EPA feels obliged, correctly many would agree, to deal with radon exposures in homes even though radon is not strictly within its statutory responsibilities. The obligation of public officials and organizations to act in the public interest even when their charge is less than perfectly clear can make the risk communicator's job more difficult. When the EPA publicizes the standards it sets, the reasons for them, and the penalties for violation for topics within its statutory responsibilities, there is little objection. But some people question the agency's justification for disseminating messages where due process has not given the agency that responsibility.

Establishing legal standing and involvement for nongovernmental individuals or organizations is more complicated. Whether a nongovernmental entity is "lawful" in the sense of corresponding to statutory requirements is less important than, for example, the concept of determination of "standing" in court, that is, of who has the right to bring suit. This concept includes notions of being materially affected by the outcome or being a representative of those who at least potentially are so affected. Being materially affected usually ensures the right to have a say in issues of public choice. In addition, personal involvement in or knowledge of the activity or events under question generally increases interest in what that individual has to say about that question. Such justification is used with respect to environmental interest groups. Finally, there is the expectation that any citizen should have the right to speak to any public issue.

Of course, being party to the creation of a hazard automatically grants standing in determining how to control that hazard. There was no question, for example, that Union Carbide should be heard from following the release of methyl isocyanate from its plant in Bhopal, India. Needless to say, however, most people expect that an organization that contributes to the creation of a hazard will make statements they believe will be in that organization's own best interest.

Justification of Communication Campaigns

People may view what it means to be a responsible communicator somewhat differently for government and nongovernmental organizations. Many people feel that government agencies should never "advocate" in the ordinary sense, that the job of public officials is to determine the factual situation, identify the impacts on affected parties, and lay out the options. The selection of what is to be done should be left to elected bodies or to the relevant individuals and organizations. In Chapter 4, however, we discussed the acceptability of influence in risk messages. We observed there that there are no clear lines distinguishing techniques that are appropriate from those that are not and that it is important to be able to demonstrate that the effort derives from a social decision supporting the communications program.

Most people would agree that an important part of the activities of federal agencies is communication about the standards they set, the reasons for them, and the penalties for violations. However, it is not always obvious when government ought to undertake programs of risk communication and how far it should go in persuading the populace to undertake particular actions. For at least some programs there are serious questions about the conditions under which persuasive risk communication should be permitted.

The central issue here is the compatibility of government-sponsored programs with individual autonomy and related values (Faden, 1987; Faden and Beauchamp, 1986). In the first place the basis for the government promoting certain life-styles over others is not clear. With major problems that affect a large portion of the populace and inflict substantial monetary and social costs, most people are prepared for government agencies to be proactive and persuasive. But exactly what makes an issue a major problem justifying advocacy on the part of government is not clear.

This, of course, is not to claim that government should never conduct risk communication programs. Some problems addressed by collective programs can be most efficiently dealt with by the affected individuals. But some question how the EPA can judge the risk of household radon exposure to be sufficient to warrant an extensive campaign of communication to homeowners, while that risk is not deemed sufficient to warrant establishment of exposure standards. One problem that may confront a government source is to justify conducting a risk communication effort rather than devoting its effort directly to reducing the risk in question. However, this can usually be justified in terms of improved efficiency in implementation of risk reduction programs.

People are generally more tolerant about communication from private enterprises or interest groups than from government agencies. These organizations are not expected to exhibit the same impartiality as government, and their attempts to present persuasive information are not viewed with the same suspicion.

Access of Affected Parties to the Decision-Making Process

Alienation of citizens due to the difficulty of getting government officials to listen to them or due to the judged inappropriateness of the officials' response has been repeatedly described as a major motivation of individuals who have become active in controversial issues. The homemaker/activist/media spokesperson appears frequently in community disputes over environmental issues (Levine, 1982; Mazur, 1987; Spain, 1984).

An important part of the legitimization of government activities that has been codified in the Administrative Procedures Act and elsewhere concerns guaranteeing the affected parties access to regulatory decisions. Government is, however, sometimes criticized for being too passive in this respect, for not actively seeking out those affected and informing them about proposed actions. This has become more obvious with respect to siting decisions and cleanup of hazardous waste facilities, where local hearings and extensive community involvement are increasingly necessary.

Nongovernmental organizations are not under the same obligation to grant the affected parties access to their decision-making processes. Nonetheless, it often appears to be advantageous to involve the public in appropriate ways. The Chemical Manufacturers Association, for example, organized a program called Community Awareness and Emergency Response (CAER), which helps chemical plant managers provide information to their communities on a regular basis and involve the community in emergency response planning, including the chemical plant emergency response plans. The CAER program, which has been implemented at more than 1500 facilities, does not suggest that citizens be involved in the actual operating decisions but rather that plant managers regularly interact with them. It emphasizes the importance of treating citizen concerns as important and providing careful and accurate responses to the public.

Fair Review of Conflicting Claims

Expectations of fair and impartial treatment by government organizations are of central importance. Nothing undermines the legitimacy of government positions more quickly than the demonstration that dispensations have been unfairly granted. It is important that all claims are genuinely listened to and treated fairly. Some criticized the EPA under Anne Burford because it allegedly attempted to rid the Scientific Advisory Board of scientists holding views of which it did not approve (Marshall, 1983b). EPA officials were also accused of inappropriately using portions of an industry publication in a "cut-and-paste" review of toxicology data (Marshall, 1983a:1200). The charge that EPA officials were not treating the issues fairly, but serving industry interests better than environmental interests, became a serious challenge to the agency's credibility (Rushefsky, 1984; Sosenko, 1983). This tarnished image caused problems for the agency for several years.

Again, there is not the same expectation that nongovernmental organizations will treat conflicts fairly. Many believe, on the basis of past experience, that industry and citizen action groups will present information that supports their interests in the most effective way. But treating the positions taken by others respectfully and thoughtfully, and carefully and clearly laying out the premises and assumptions of one's position, are likely to enhance the reputation of nongovernmental organizations as well.

Making Messages Understandable

The risk communicator needs to present information in language and concepts that recipients already understand, that use magnitudes common in ordinary experience, and that are sensitive to the psychological needs of the recipients.

Unfamiliar Language

In Chapter 2 we described the scientific information needed for risk decisions and the difficulty of presenting that information in simple terms that do not overwhelm the recipient. Here we will point out some of the ways the terminology of science, and of risk assessment in particular, interferes with understanding by laypeople.

For those who are not familiar with it, the technical terminology of risk assessment is very difficult to understand. Research has shown, for example, that even for seemingly familiar terms such as probabilistic precipitation forecasts there is a high degree of misunderstanding in lay interpretations. People were equally likely to interpret a "70% chance of rain" as "rain 70% of the time," "rain over 70% of the area," and "70% chance of some measurable rain" (the official definition) (Murphy et al., 1980).

In this case people apparently had difficulty understanding not the probabilities being used but the events to which they were applied. In other cases risk assessments confuse people because they use concepts of probability theory that are not intuitive. For example, a committed communicator can usually convey the meaning of a simple probability of an event occurring (as long as it is not too small). But the notion of conditional probability is much more difficult to get across, as is that of the probability of the conjunction of several events. In the EPA's experience with EDB, people were confused by the notion of the aggregate risk of this pesticide to the exposed population. They did not know how to interpret this information to answer the question, "Should I eat the bread?" (Sharlin, 1987).

Much has been written about the intuitive properties of probabilistic events and how they differ from probability theory (see the summary in Fischhoff et al., 1981a). These include the "gambler's fallacy" (after a series of heads in flips of a fair coin, most people expect a tail) and the tendency to impose order on the results of random processes. The skeptical recipient of risk messages will be on the lookout for such influences in the material he or she receives.

Recipients of risk messages also need to be wary of "framing effects"—differences that can result from the way information is presented. For example, one guide for chemical plant managers points out the following ways that the annual fatalities resulting from emission of an air toxic might be presented (Covello et al., 1988):

- deaths per million people in the population,
- deaths per million people within miles of the facility,
- deaths per unit of concentration,
- deaths per facility,
- deaths per ton of the airborne toxic substance released,
- deaths per ton of the airborne toxic substance absorbed by people,
- deaths per ton of chemical produced, and
- deaths per million dollars of product produced.

The authors point out that depending on the circumstances different expressions will strike the recipients as more or less appropriate, more or less frightening, or more or less credible. The recipient needs to be aware that simply changing the way a piece of information is presented can alter its effect on many people and be aware of their pattern of response. Each of these ways of presenting has framing effects, which suggests that using more than one might be useful in some circumstances.

Misunderstanding can also result from inconsistency in the use of the same term among different disciplines. For example, the term "risk" has been used with a variety of somewhat different meanings. Uses include the total number of deaths, deaths per person exposed or per hour of exposure, loss of life expectancy due to exposure, and loss of the ability to work (Fischhoff et al., 1986).

The recipient also needs to look for the sources of uncertainty in the analysis. As we pointed out in Chapter 2, at least four types of uncertainty may be found in risk messages: (1) weaknesses of the available data, (2) assumptions and models on which estimates are based when data are missing or uncertain, (3) sensitivity of the estimates to changes in the assumptions or models, and (4) sensitivity of the decision to changes in the estimates.

Unfamiliar Magnitudes

Risk communicators are generally well aware that most people have difficulty comprehending magnitudes that are exceedingly small or exceedingly large. Often they utilize analogies to convey such magnitudes. For example, a risk of 0.05 may not mean much to most people; the statement that about 5 people in an auditorium of 100 would be affected is much easier to comprehend. A cancer risk of 4.7×10^{-6} is difficult for most people to relate to. But it may be more understandable to imagine 10 cities of 100,000 people each, all exposed to the hazard. In 5 of the cities there would be no effect, and in 5 cities there would be 1 additional cancer as a result of the exposure (Covello et al., 1988). Such mental aids must be used with extreme care, however. For instance, in this example there could be 5 additional cancers in a single city instead of spread across the cities, and there could be an overall total of 10 or of 0. Recipients need to look for the magnitudes in risk messages and how they are presented. These numbers are subject to the same kinds of presentation effects as the concepts described in the previous section.

Recipients need to be especially wary of misrepresentation that can be introduced in comparing risks. In particular, magnitudes do not always represent the level of hazard. For example, characterizing the magnitude of ash emitted as filling an olympic-sized swimming pool or covering a football field to a depth of 6 inches omits any reference to the potency of the material. Because of differences in potency, a fairly small amount of one substance may present the same risk as a much larger amount of another.

Insensitivity to Psychological Needs of the Recipient

Factors influencing understandability intermingle with certain psychological components of decision making. Motivational factors can be involved. For example, it may be difficult to determine whether the intended recipients have not understood the message or whether they have understood but decided for whatever reason not to heed its contents. People may want simple yes-or-no answers, and they may want to know what they as individuals should do. When expecting information in such a format, they are likely to have trouble understanding information presented in some other format. The first image people receive about a problem also tends to be the strongest and longest lasting. If they make up their minds on the basis of that image, it will be hard to get them to change. But the risk communicator can use such psychological attributes to his or her advantage, as well. If the risk communicator is timely and presents a vivid image, he or she can have considerable impact.

With certain issues and certain parts of the population, communication may be especially difficult. There may be, for example, a climate of mistrust in parts of the population about anything that can be labeled toxic. These people may automatically reject a message and oppose the production, use, or disposal of products labeled toxic regardless of the risk estimates of experts. For them it may be that risk messages would elicit little differentiation of response regardless of their format, message content, or the organization from which they emanate.

People are unlikely to be interested in risk information that they cannot use. A risk communicator wishing to change the recipient's thinking (even if only to make him or her better informed) thus needs to try to understand how that person receives, processes, and acts on information. Elsewhere we discussed the psychology of risk perception and social factors that influence perceptions of risk. Here we review only the more important of those psychological and social influences.

People differ. Their interests, life-styles, and living conditions vary. What they do in their private lives and how they interact with others in their public lives will strongly influence how they are likely to use risk information. The risk communicator will be most effective if these attributes of recipients can be reflected in the risk message.

For issues that affect large numbers of people, it will nearly always be a mistake to assume that the people involved are a homogeneous group. It is therefore generally necessary to segment the population into groups with similar needs. It is often useful to craft separate messages that are appropriate for each segment. Preparing messages appropriate for different segments of the population requires determining what the recipients already know or think they know, what is necessary for a full and sufficient understanding of the risk and risk reduction measures, and how they would be able to use new information. Depending on the numbers involved, this can be expensive and time consuming.

The purpose the message serves can dominate people's information needs and therefore the content of an effective risk message. For example, information needs will probably be quite different when the situation calls for providing emergency instructions, for alerting people to a previously unrecognized risk, or for providing information that a risk is actually less serious than was previously thought.

A common mistake is to expect quantitative risk assessments to include everything people are concerned about. Affective states (those involving or appealing to emotion) are equally or more important than physical conditions to many people. Since risk assessments are usually limited to physical events and consequences, they can be expected to speak to only part of what concerns most people. These other aspects of risks that concern people are sometimes called qualitative risk factors.

Thus, in order to present information that is relevant to the intended recipients, it may be necessary to expend some effort to find out what is bothering people. To be effective, a risk message needs to refer both to information about risk and risk reduction and to the psychological or affective factors that influence the intended recipients. Unfortunately, it can be difficult and expensive to develop empirical information about recipients, especially if they are geographically or culturally dispersed.

It is important not to expect too much from risk communication efforts. Advertising campaigns are considered successful if they result in shifts of a few percentage points in the market for a product. It took decades of multiple messages from many different sources to create major shifts in public attitudes about smoking. It is hard enough to make risk messages understandable to laypeople. It is harder still to know whether risk messages have an impact on their thinking.

Preparing Messages with Few Data and No Time

Sometimes the risk communicator must disseminate messages when there are not enough relevant data to draw satisfactory conclusions and there is no time to obtain better information. This usually occurs in one of the following situations: (1) an emergency requires immediate action or (2) events lead to requests for information prior to the completion of study or analysis.

Responding in an Emergency

Emergencies occur when external events take control and require action by an individual or organization. They often, but not always, require immediate issuance of warnings, instructions, or advice about what to do. Examples of such emergencies include Three Mile Island, the Tylenol poisonings, and emergency releases from chemical or other industrial facilities.

The problem is most extreme in a true emergency when no preparation has been made in advance of the event. For example, the Nuclear Regulatory Commission was almost totally unprepared for an accident at the time of Three Mile Island (Ahearne, personal communication with National Research Council staff, 1988). There was no effective management structure to support emergency decision making, and time was lost figuring out who should do what. The lack of preparedness permitted the involvement of too many who lacked the technical competence to grapple with the emergency, thereby slowing the rate at which necessary information could be generated and interpreted. No one with the technical background to explain what was happening had been assigned the role of spokesperson, and it was a couple of days before a credible source of information emerged. Nor did the agency, much less the electric power company involved, appreciate the importance of timely and accurate news releases. Finally, the agency had no notion of how to deal with the electric company or the news media in such an emergency.

Emergency situations are likely to expose the risk communicator to conflicting motivations. For example, a company dealing with an emergency release of toxic substances into the air, such as that from the Union Carbide plant at Institute, West Virginia, in August 1985, will probably balance several competing factors in deciding what messages to give out (Coppock, 1987). After the initial emergency response, when the overriding concern is what to do to contain and stop the release, almost every business person immediately wonders who will bring suit. The common view among legal advisors is almost always to give out as little information as possible so as to avoid providing ammunition for use in court. This is in almost direct conflict with what communications and community relations experts advise, which is to say everything that is known, as quickly as possible, in terms the layperson can easily understand. Advice of company scientists and engineers usually falls somewhere between these two views. They caution against attributing cause and effect before being reasonably certain about what happened. The message that is finally sent out probably involves compromises between these three points of view.

Situations involving emergency response are often governed by special considerations not shared with other kinds of risk communication. We have chosen to focus primarily on the other, more prevalent situations of risk communication and therefore do not discuss emergency response in detail.

Communicating on the Basis of Incomplete Information

It is often difficult to estimate risks, consequences, and possible risk reduction measures with any precision. One result is that the risk communicator may be left with very little information that can be presented with confidence. As one scientist at the EPA put it, "One of the nice things about the environmental standard setting business is that you are always setting the standard at a level where the data is lousy" (quoted by Melnick, 1983:244).

The poor quality of relevant information is also often involved in pressing issues. When the concern about EDB shifted from groundwater contamination in a few isolated wells to residues in food products, EPA administrator William D. Ruckelshaus sent a letter to the governors of the 50 states requesting data on residue levels in food products. He had to answer queries by admitting that his agency did not have the answers. "If they [the public] want absolute information, we can't give it to them." For a period of nearly a month, the best he could do was say, "I don't want to unduly alarm the public, nor do I want them not to know about it" (Sharlin, 1987:192). The risk communicator may often feel as if the world wants to know definitive answers to questions about which he or she has no adequate information.

External demands can also force an organization to make statements on the basis of limited data. Examples include Love Canal and transmission of the AIDS virus. Another form in which this problem can be found is the decision about whether to release preliminary information or tentative results. In 1986 the EPA began cooperating with the New York State Energy Research and Development Authority (NYSERDA) on a program monitoring radon levels in geographic regions thought to have radon problems (Smith et al., 1987). Three monitors were placed in each home, one in the basement for 2 to 3 months, a second in the basement for 12 months, and a third in the living area for 12 months. Originally, the plan was to give the homeowners the readings from all three monitors at the end of the study. But in the spring of 1986, when radon became very much a public issue, NYSERDA became concerned that they would be accused of withholding public health information if they kept the short-term reading until the end of the study. It was decided to provide the initial basement reading to the homeowners, even though the full research design called for confirmation of annual exposure levels and living-area exposures with the other monitors (Fisher, 1987).

Capturing and Focusing Attention

Many other things compete with risk messages for attention, and the risk communicator often has difficulty getting intended recipients to attend to the issues. There are two separate aspects to this problem: (1) stimulating the attention of recipients and (2) interacting with the news media and other intermediaries.

Stimulating Recipient Interest

It is not always easy to capture the attention of people who receive risk messages. Most information campaigns share the following attribute: the people most likely to receive messages and to attend to them are those who already possess some information about the issues under question; those who may be characterized as relatively uninformed are less likely to receive and pay attention to messages. The people who need information most seem to be the least likely to pay attention. One contemporary example might be the very low likelihood that intravenous drug users will attend to messages about AIDS transmission via dirty hypodermic needles.

Involvement in community affairs has been characterized as a pyramid. At the bottom is the broad base of most people who are uninvolved in any personal sense and basically are uninterested. A somewhat smaller number of people are aware of issues but do not go to much effort to obtain additional information. A still smaller group actively seek information on particular issues. The number of people actually participating in organized efforts is smaller still. Finally, some individuals seem to participate in, and often lead, every activity in a community (Verba and Nie, 1972).

One of the consequences of this differential interest and involvement among various parts of the public is that information readily made available will generally be taken up much more readily by some people than by others. Educated and involved people usually absorb information much more quickly. But conflict can motivate otherwise uninterested people to gain more information. In some situations, when the principal aim is to stimulate understanding in a broader sector of the public, it might be useful to stimulate conflict. Very often journalists and the media seek out such conflicts and serve as information channels as conflicts play out.

The majority of people, however, will probably not be interested in the issues addressed in a particular risk message. When a significant number of people are similarly affected, a champion for that group is likely to emerge, especially when the impact is undesirable. Such people can be engaged in risk communication activities as described elsewhere in this report. When trying to affect the behavior of uninterested, uninvolved people, however, the risk communicator will need to find ways of attracting the attention of intended recipients and making the message meaningful to those people. This will probably be easier if the risk is one that is perceived to directly impinge upon people and for which there are clear control measures that do not substantially interrupt their private lives. For example, people have tended not to heed messages about seat belt use, maintenance of automotive emission control devices, and radon contamination of homes. However, it is difficult to determine whether they simply paid no attention or whether they received the information, understood it, and decided not to act in accordance with the proffered advice.

Different people rely on different information channels. They read different newspapers and magazines and listen to different radio and television stations. They may turn to different information channels for different purposes. Young people, for example, may rely on mass media sources to learn about the AIDS epidemic and its spread. But they may turn to their friends in determining whether to be worried and alter their behavior. Risk communicators need to know what channels their intended audience uses for what aspects of risk information. One example of this is the use of music television spots by the National Cancer Institute to convey the message to teens that it is not "sexy" to smoke, rather than providing information about the undesirable health effects of cigarette smoke.[1]

Interacting with the News Media and Other Intermediaries

The mass media are widely perceived as playing a powerful role in constructing laypeople's understanding of and attitudes about risk. Journalists and the media help identify conflicts about risk and are important channels of information during the resolution of those conflicts. There are both critics and defenders of the effects of the news media. In any case the risk communicator must deal with the fact that some journalists tend to treat risk issues differently from the way technical and scientific people do.

Some conflict between risk communicators and journalists and other intermediaries is probably inevitable. But this conflict can be reduced, and there are approaches the risk communicator can use toward achieving this aim. An important part of this is to recognize the typical differences in the way risk communicators, as sources of information, and journalists approach information gathering and dissemination.

Organizations involved in risk issues typically seek to centralize and restrict the flow of information, hoping to prevent the publication of damaging information. But reporters expect access not only to public information specialists but also to experts and managers and what they know (Sandman et al., 1987a). This is especially true in emergency situations. The price of not providing that access may include suspicion, anger, and sometimes damaging coverage. Despite the legal and technical constraints, it is important to consider meeting the needs of the news media.

Many journalists are proud of their ability to flesh out a story with the views of uninvolved experts, dissident insiders, and others whose perspective on an event is likely to be different from the official one (Sandman et al., 1987a). Specialized reporters are proud of their contacts and investigative reporters of their skill at finding those who know and of persuading them to talk. Trying to stop reporters from talking to people within an organization is sure to encourage them to investigate further.

Differences of opinion as to what should ideally be presented are likely to exist between risk communicators and journalists. Sources sacrifice all credibility in the eyes of reporters when they lie or mislead, and they lose much of it when they err, omit, or delay (Sandman et al., 1987a). Different sources are commonly held to different standards of credibility. Industry spokespeople, for example, are often discounted as opinionated even when they are providing facts, while academics and public interest groups often are accorded the credibility of neutral sources even when they are offering opinions.

Journalists, too, have a problem with credibility (Sandman et al., 1987a). A botched story not only misleads the reader or viewer but it also diminishes the source's willingness to cooperate with that reporter next time and perhaps with other reporters as well. The two most important complaints about reporters' treatments are misquotation and inaccuracy. Technical stories have greater chance of misquotation, simply because they involve terms and concepts less familiar to the reporter than nontechnical stories. Nevertheless, incompleteness and misemphasis—quoting out of context—are more frequent than direct misquotation. Complaints about inaccuracy are also generally about being incomplete or misleading. Sometimes the complaint is that too much credence is given to other sources who, in that source's judgment, are wrong or intentionally misleading the journalist. These questions are commonly sources of conflict between risk communicators and journalists, especially because the journalist does not see his job as discovering the truth, but rather as reporting accurately what others with some claim to attention consider to be true.

Risk messages are often routed to their intended recipients through health professionals or other intermediaries. In addition, the views of influential members of the community, such as county or local public health officers, prominent physicians, fire chiefs, and politicians, often provide valued guidance to citizens as they form their opinions about controversial issues. Sometimes executive officers of professional or volunteer organizations serve as "gatekeepers," controlling the distribution of information, and their approval or disapproval can be a critical factor in the dissemination of some risk messages. In some circumstances the intermediaries are even more important than the news media in reaching the intended recipients.

Interacting with non-news-media intermediaries can also involve problems. Health departments, public libraries, professional associations, and voluntary organizations all have their own aims and purposes. They may or may not offer relevant messages to the intended recipients that are appropriate in terms of format and style. It may be quite time consuming and costly to establish working relationships with such intermediaries, however, and there is the danger of losing control over the content of the messages. Nevertheless, establishing links with such institutions and organizations can shortcut the development of routes of influence with the target recipients.

In this context it is important to realize that there are several different ways that messages can reach the final recipients: face to face (physician to patient, friend to friend, within the family), in groups (work sites, classrooms), within organizations (professional or volunteer), through the mass media (radio, television, magazines, newspapers, direct mail, billboards, and transit cards), and within the community (libraries, malls, fairs, and local government). Each of these channels offers advantages and disadvantages in specific situations.

Interpersonal channels like physicians or pharmacists are likely to be trusted and influential. But messages relying on interpersonal channels require the intermediaries to be thoroughly familiar with the message and may thus require expensive and slow long-term contact.

Community channels such as libraries and community organizations can reinforce and expand upon media messages. Establishing links with community organizations can require less time than reliance on interpersonal channels.

Using celebrities can be effective if they are directly associated with the message (e.g., they have been a cancer patient, are pregnant, or successfully altered a hazardous habit). But they speak for themselves, and it is important to have firm agreement about what they will—and will not—say. The appearance of a celebrity may compete with the content of the message for attention, and some recipients may not react favorably to some celebrities. Finally, celebrities live in the public eye and a change in their popularity or personal life style could affect their impact.

Working with intermediaries is essential in many situations. Intermediaries can help by providing special access to the intended recipients, credibility because the recipients consider them to be a trusted source of information, and additional tangible or intangible resources. Working with these individuals and organizations, however, can also have drawbacks. It can be time consuming to locate them, convince them to participate, gain their approval, and develop and agree on their role. It can require adjustment in order to match the priorities and programs of intermediary organizations. It can result in loss of control of the risk message because they may change the time schedule, functions, or even the content of messages and take credit for part or all of the effort.

Getting Information

Recipients of risk messages may have difficulty deciding what to do because they cannot get information that satisfactorily answers their questions. This can result from one or both of the following: authorities who do not listen or who respond inappropriately and difficulty in finding trusted local sources of information.

Authorities Who Do Not Listen or Respond

The story of the concerned citizen motivated to organize protest groups because of the cold or indifferent response of public officials is common in the literature of environmental and citizens' organizations (Fitchen et al., 1987; Institute for Environmental Negotiation, 1984; Krimsky and Plough, 1988; Mazur, 1987). A citizen who had spent several years as an activist opposing the construction of a hazardous waste facility in her community told us of the frustration her group experienced in trying to get the authorities to take their concerns seriously and in attempting to obtain materials they could use to inform themselves and their neighbors (Smith, 1987). She spoke of the anger generated by the lack of respect given her group's questions by government officials. At a public hearing the company proposing to construct the facility was allowed to speak freely. But questions from the public had to be submitted in writing. Nor was the citizens' group able to find support from the traditional national environmental organizations. Finally, they turned to other citizens' groups who were opposing the same company in other locations. This may be a common experience for citizens' groups focusing on locally unwanted land uses. The number of Superfund sites around the country and the pressing necessity for finding ways of dealing with hazardous wastes will make this kind of difficulty likely to reappear many times.

Difficulties in Finding Trusted Sources of Information

Other developments will result in citizens or citizens' groups seeking additional information. For example, Title III of the Superfund Amendments and Reauthorization Act of 1986, also called the Community Right-to-Know Act, includes provisions for creating emergency response plans and for reporting data about hazardous substances stored and regular emissions to the EPA. The EPA must make these data available to the public. The Community Right-to-Know Act will make a tremendous amount of information about potential hazardous situations available to citizens who wish to obtain it. But this information is likely to be in highly technical form, most of which would require considerable interpretation to appreciate. A citizen wishing to make sense of this information about a facility in his or her community will need to interpret data from material safety data sheets developed for occupational exposures and estimate peak or periodic exposures from annual emission totals. He or she may wish to seek additional interpretations to those provided by facility personnel, and finding trusted and qualified people to interpret this information will be an important part of the process.

SUMMARY

We distinguish two major types of problems in risk communication. Those involved in risk communication can do little about problems deriving from the institutional and political system beyond understanding them and their influence. These problems can have considerable impact on events, and if they are ignored it may be quite difficult to understand why things happen the way they do. Problems of risk communicators and recipients can be addressed more directly and are more amenable to improvement or solution. In most instances the problems of risk communicators and the recipients of risk messages are mirror images of each other. In the next chapter we describe conclusions and recommendations that are intended to improve risk communication in ways that will address the problems of risk communicators and of the recipients of risk messages.

NOTE

1. This is unlikely to meet our criteria of informing or of accuracy of the message.

7
Recommendations for Improving Risk Communication

Drawing lessons from the available understanding about the nature and problems of risk communication, we present four sets of recommendations in this chapter: (1) recommendations that pertain to the processes that source organizations use to generate decisions, knowledge, and risk messages; (2) recommendations that pertain to the content of individual risk messages; (3) a call for a "consumer's guide" that will enhance the ability of other groups or individuals to understand and participate in risk management activities; and (4) a brief summary of particular areas for which additional knowledge is needed to resolve current problems of risk communication.

We have attempted a focused search. The committee faced a central dilemma about how detailed we could expect to be in meeting our charge to discern practical lessons for practitioners. Given the breadth and diversity of the general topic of risk communication, any attempt to look for lessons that apply to all forms of risk communication would constrain us to a discussion so general that any particular reader would gain little insight. On the other hand, a detailed "cookbook" for particular situations would fail to advance the broad national discussion that is now needed. We have accordingly sought a middle ground, electing to narrow our scope in two ways.

First, we have elected to focus on certain forms of risk communication. The term "risk communication" can cover a vast range of actions, from casual telephone calls between two experts to book-length reports meant for the general public. Our main subject in this chapter is formal risk messages intended for audiences that include nonexperts. Included, for example, are press releases, material prepared for an open meeting in a community or a formal meeting with representatives of interested outside groups (e.g., a local public meeting about siting a facility), a government agency's public explanation of a decision it has made, a brochure for citizens concern-some aspect of public health (e.g., an AIDS pamphlet), package inserts for prescription drugs, and risk summaries prepared by experts within an organization for the use of their (less-expert) superiors. We recognize that some of our recommendations may have less relevance for other very important, but less formal, varieties of risk communication.

Second, we have directed our recommendations to just two of the many types of risk-managing organizations that are discussed in other parts of our report: namely, government agencies and large private corporations. Again, this choice of emphasis is not intended to imply that other communicating organizations and individuals—small firms, citizen/consumer advocacy groups, and so on—are not important. In fact, many of the points we raise doubtless apply to them. We chose this narrower range of organizations because they are most directly involved in many of the best known and most controversial cases, the committee members have greater knowledge of their experiences, and we are convinced that improvements by these organizations would both contribute substantially to easing the national problem and provide models for other organizations.

Our objective, then, is to improve risk communication, particularly as practiced by government and large corporations. What do we mean by "improve"? We mean that solutions—sometimes admittedly only partial solutions—are put in place for the range of problems identified in the previous chapter. We emphasize in particular that we have tried to fashion recommendations that, while addressed to government and large corporations, will attack the problems of recipients as well. Our goal is not then to make those who disseminate formal risk messages simply more effective by improving their credibility, understandability, and so on—such an approach might serve their interests but could well degrade the overall quality of risk communication if it meant that they would merely advance their viewpoints with more influence. "Improvement" can only occur if recipients are also enabled to solve their problems at the same time. Generally, this means obtaining relevant information for better-informed decisions.

We have also focused our recommendations on measures that will help those groups meet the criteria we have set out above for successful risk communication. In reality, of course, many organizations have other criteria for success, such as whether messages convince recipients to act in a manner that the risk communicator desires. We have not chosen to recommend actions to help organizations meet those other goals.

In recommending steps to be taken by government entities, we have necessarily focused on the respective roles of citizens, private groups, and government in a democratic society. Controversies about risk communication often turn out to be basic debates about the limits of governmental accountability, legitimacy, and authority. The goal of our recommendations is not to alter American democratic institutions but to make them work more effectively. Two points need to be emphasized about accountability. First, our society has elaborate and politically responsive procedures for assigning responsibilities for making government risk management decisions. Once a government agency has received that responsibility, it must retain it. This places inherent limits on what agencies can do in discussing risk issues with citizens, because they cannot share responsibility with outside groups; they must remain publicly accountable. Second, accountability increasingly implies an affirmative duty to interact with interested and affected outside parties in reaching and explaining individual policy decisions. Although citizens—and the groups that undertake to represent their interests—are not required to participate in such interactions, solving problems of risk communication becomes much easier if they do, and government needs to ensure that the opportunity to participate becomes routine.

Implementation of many of our recommendations requires organizational resources of several kinds. We are aware that such resources will not be adequate in many instances. One resource in particular—time—is crucially lacking for some of the most difficult risk communication efforts, as when emergency conditions leave no possibility of consulting with outside organizations or assembling complete factual information. Other recommendations require staff resources and the capacity to conduct specialized analyses, both of which may be in short supply in some organizations. When resources are so constrained, our recommendations may well best serve as a reminder of the full set of factors that should be accommodated, although the form of accommodation may fall short of what we recommend.

Our recommendations are based on our understanding of the growing literature of studies of risk communication and risk messages and on committee members' diverse experience with specific instances of risk communication.

Before we list our recommendations, we would like to draw attention to three general conclusions that we have made:

Conclusion 1. *Even great improvement in risk communication will not resolve risk management problems and end controversy (although poor risk communication can create them). Because risk communication is so tightly linked to the management of risks, solutions to the problems of risk communication often entail changes in risk management and risk analysis. There is, unfortunately, no ready shortcut to improving the nation's risk communication efforts. The needed improvement in performance can only come incrementally and only from assiduous attention to many details.*

While it is important to improve risk communication practices, no one should expect such improvements to end public controversy over risk management. Risk managers should understand and accept that, even when they have done all they can to ensure the integrity of their risk messages, public skepticism of their motives and their honesty will likely persist. They should appreciate that, particularly in recent years, distrust has been institutionalized in our country. While it is important for most risk managers—especially those in the government—to avoid distortions in their messages, they should expect that many audiences will continue to assume that bias is present.

We have discovered no sweeping broad-spectrum remedies for the problems of risk communication described in Chapter 6. Many will be solved only over the long term and only by sustained effort. Many of the institutional problems we identified in the previous chapter—fragmentation of authority, legal constraints, and so on—reflect social decisions about how risk management should be conducted. Such decisions are inherently, and appropriately, political in nature. Risk communication might well be improved if certain contextual constraints were changed or removed. However, such reforms would also create other advantages and disadvantages that are well beyond our capacity to evaluate in this study. Thus we are left with a more modest, and necessarily incremental, set of available remedies.

The source organization's problem of achieving credibility provides a good example. An organization's credibility can be quickly lost, as illustrated in the case of the EPA in the early 1980s, when many observers came to believe that one of EPA's leaders' highest goals was to dismantle regulatory programs. In contrast, credibility is gained (or regained) only through a sustained effort to be responsive to audience concerns and to be accurate, open, and honest in disclosing essential information. Thus we are led to recommend concurrent attention to several factors in managing the risk communication process and in formulating particular risk messages. No one of these measures, alone, is enough.

An underlying reason for this is that the problems of risk communication are rooted in risk management practices and procedures. Because of this, several of the measures we recommend call for adjustments in the source organization's procedures for risk management and for analyzing risk issues. For example, we call for more interaction with audiences and intermediaries while the source organization considers risk management alternatives, and we suggest how formal risk assessments should be scoped, reviewed, and presented. We have explicitly addressed many of our recommendations to risk managers precisely because they are the individuals within an organization who can provide the needed coordination of risk communication, risk management, and the assessment of risk and risk control.

Conclusion 2. *Solving the problems of risk communication is as much about improving procedures as improving content. Risk managers need to consider risk communication as an important and integral aspect of risk management. In some instances, risk communication will, in fact, change the risk management process itself.*

It would be a mistake to believe that better risk communication is mainly a matter of crafting better messages. To enhance credibility, to ensure accuracy, to understand recipients and their concerns, and to gain the necessary insight into how messages are actually apprehended, one must ultimately seek procedural solutions. Thus we devote much of this chapter to matters of process. There may be many cases in which problems of credibility, potential controversy over value judgments, and diverse audiences reduce the risk communication task to a simpler matter of making messages clearer, in themselves. We do not believe that the national frustration over risk communication practices derives from failures in such "simpler" cases and therefore have not addressed simpler cases in any detail.

Risk managers cannot afford to treat risk communication as an afterthought. One of the root problems in risk communication is that, perhaps due to organizational imperatives and tradition, risk management has too often been treated as a sequential process: (1) the organization's technical experts assess a risk and explore options, (2) a risk management decision is made, (3) a message is internally prepared, and (4) the message is sent to outsiders. Risk communication is thus regarded as a subsidiary activity.

The importance of risk communication has only recently become apparent, and even the most progressive risk managers are only now beginning to adjust to the realization. Improvement of risk communication requires that the organizations that disseminate risk messages become simply more deliberate in their communication efforts.

At their best, risk communication efforts can be expected to affect the risk management process itself. Considerations of risk communication might, for example, determine what kinds of analyses of risks and benefits are performed, how risk assessments are summarized, what options are explored, and what people are consulted in exploring possible courses of action.

Risk communication requires its own specialized expertise and deliberate planning and evaluation. Senior managers need to devote attention and time to managing risk communication efforts per se. It is a mistake to simply consider risk communication to be an add-on activity for either scientific or public affairs staffs; both elements should be involved. There are clear dangers if risk messages are formulated ad hoc by public relations personnel in isolation from available technical expertise; neither can they be prepared by risk analysts as a casual extension of their analytic duties.

Conclusion 3. *Two broad themes are apparent in the extended list of recommendations: that communication efforts should be more systematically oriented to specified audiences and that openness is the surest policy.*

Both the management of the process of formulating risk messages and the content of risk messages should be systematically oriented to the intended audience. The most effective risk messages are those that quite self-consciously address the audience's perspectives and concerns. Similarly, the best procedures for formulating risk messages have been those that involved interactions with recipients and that elicited recipients' perceptions and needs.

A central premise of democratic government—the existence of an informed electorate—implies a free flow of information. Suppression of relevant information is not only wrong but is usually, over the longer term, also ineffective. Risk messages should be explicit about current knowledge of the subject risk but also about the limits of that knowledge and the existence of disagreement among the experts or others. The long-term improvement of credibility, in particular, depends on openness. Several of our procedural and content recommendations are intended to foster openness and to promote openmindedness about outside viewpoints.

MANAGEMENT OF THE PROCESS

Much recent concern about risk communication has centered on questions of message content. Failures have frequently been attributed to the inability of the audience to comprehend complex technical issues and to the tendency of risk messages to be badly written. This view would lead one to seek solutions in the design of better risk messages themselves. Our assessment has led us to believe that longer-term solutions are equally likely to involve attention to and changes in the process by which risk management decisions are made and explained.

There are two basic reasons for our emphasis on process. First, when lessons about message content are identified, the operational question becomes one of ensuring that those lessons are systematically followed. Procedural safeguards provide the best assurance of routine compliance. Second, and more important, it is increasingly clear that content and process are not easily separated, particularly on the crucial matter of appearing credible. If recipients believe the process is flawed—for example, if the communicating organization is known to ignore or reject certain facts, viewpoints, or options—they are likely to doubt the message, even if it is, in fact, technically competent.

This section is addressed to risk managers—those senior officials who have the overall responsibility of determining their organization's action. These risk managers also oversee the preparation of risk assessments and risk messages associated with the action to be taken.

We identify four process objectives that are key elements in improving risk communication: goal setting, openness, balance, and competence. We note that these objectives are general in nature.

Different management styles may work best for different managers in particular situations, in pursuit of these four objectives.

Setting Realistic Goals

Some past deficiencies in risk communication efforts have arisen because risk managers have not appreciated that risk communication needs to receive deliberate management attention. Until now, risk communication efforts have all too often been pursued with implicit or impractical objectives within the source organization.

Risk communication activities ought to be matters of conscious design. Practical goals should be established that explicitly accommodate the political/legal mandates and constraints bounding the process and the roles of the potential recipients of the organization's risk messages. Explicit consideration of such factors encourages realistic expectations, clarification of motives and objectives (both within the source organization and among outside groups and individuals), and evaluation of performance.

Consideration of these issues of practical goals and impediments to their achievement may be the only way for managers to reach realistic expectations. Otherwise, source organizations may set themselves up for frustration and, if naive or insensitive programs result, for disrespect among recipients that can only aggravate any preexisting tensions about how the risk should be managed.

Effective program management is enhanced by setting explicit objectives. This is especially important with respect to risk communication because of the difficulty of assessing the effect of messages. A cornerstone of systematic risk communication goals is a realistic review of the political and legal context of the communication effort and the risk management decisions to which it relates. What is one empowered to do? Can messages properly attempt to induce recipients to take certain actions or can they only transmit neutral information? Who must receive the information? What level of understanding (if any) must be assured? How active a part can interested and affected parties be allowed to play in the risk management process? Analysis contributing to goal setting provides a way to articulate the basic premises for action and a basis for evaluation of performance.

Such analysis sets the general context for a risk communication effort. It needs then to be translated into operational objectives. For example, how many people should receive the message? What changes (if any) should be observed in recipients' beliefs or actions regarding the risk? Will recipients be motivated to listen? Will they rely on other, possibly contradictory, sources? Realistic assessment of factors affecting message preparation, transmission, and receipt can be an important contribution to an organization's effective participation in the risk communication process.

Safeguarding Openness

In many cases risk communication efforts have foundered because public trust and credibility were damaged because risk management was conducted behind closed doors or because of a patronizing attitude toward interested outside groups.

Risk communication should be a two-way street. Organizations that communicate risks should ensure effective dialogue with potentially affected outsiders. This two-way process should exhibit:

• a spirit of open exchange in a common undertaking, not a series of "canned" briefings—discussion should not be restricted to technical "nonemotional" issues—and

• early and sustained interchange, including the media and other message intermediaries.

Openness does not ordinarily, however, imply empowerment to determine the host organization's risk management decisions. To avoid misunderstanding the limits of participation should be made clear from the outset.

Risk managers should resist the temptation to close their processes to outside scrutiny and participation unless, as is rarely the case, extreme conditions warrant secrecy. As a practical matter, problems of risk communication for many past cases seem most pronounced when risk communicators have not appeared to value openness. In addition, many of the cases that were resolved relatively effectively were marked by openness.

Openness thus has practical benefits both for the organization that manages risk and for outside participants, but there are deeper reasons for it. Openness is highly valued in a democratic society like ours because public accountability is a central element of our political culture. This is particularly true for organizations that are responsible to an electorate or that are charged with a public purpose, but private organizations are hardly immune in contemporary America. The fact that ours is a democratic culture means that there are strong negative sanctions in public opinion for evidence of secrecy. When governments or corporations can be found guilty of withholding information, they commonly find themselves severely condemned, and their credibility is damaged for some time, regardless of the content of their risk message. Thus openness may be seen both as a matter of principle and a matter of practical wisdom for operating in a culture where many others take openness to be a matter of principle.

Openness may take diverse forms in diverse risk management settings. When a government agency considers issuing a regulation, it can involve representatives of interested and affected groups in discussions of the rationale for action, quantitative and qualitative indications of the subject risk, available alternatives, and other factors affecting its choice. If an organization undertakes to advise the general public of a risk associated with personal behavior (e.g., diet, sex), it can involve representatives of the intended audiences in discussions of the need for risk messages and the best ways to compose them. If a corporation decides to locate a new facility in a community, it can draw community groups into discussions of the nature of risks presented by the facility and take steps to control such risks. Risk messages will prove much more difficult to convey when recipients believe they were excluded from risk management decisions that affect them.

Openness also provides an opportunity for risk managers to receive important information from outside the organization relevant to their risk management decisions, as is amplified in the later discussion of competence.

Effective Dialogue

The most productive interactions are those that treat outside parties as fully legitimate participants, so that two-way exchange occurs. If the host organization conveys the impression that it is meeting with groups simply to diffuse outside concerns, or to edify "uninformed" lay risk perceptions, this goal cannot be met. If mutual trust is established, the host organization will benefit from fresh ideas, will understand better how its formal risk messages will be perceived, and will be able to incorporate needed adjustments to messages earlier than if opposition forms in response to a message. Participating organizations will have a chance to understand the basis for action and to determine for themselves the degree to which the risk decision and the associated risk message are based on full and open-minded consideration of available knowledge and the full range of alternative actions.

Eliciting participation is not simply the passive provision of access to the process of forming risk messages. Many outside groups have had frustrating experiences in which their views have been elicited but not listened to. An example is the holding of pro forma public hearings, which frustrated participants later feel should have been labeled as "talkings," not hearings, from the host organization's apparent lack of attention to points raised. Active effort should be applied to identifying the full set of interested and affected groups and ensuring that the full range of potentially contending viewpoints is apprehended. The risk manager should ensure that those in the organization have come to understand:

- what the participants know, believe, and do not believe about the subject risk and ways to control it;
- what quantitative and qualitative information participants need to know to make critical decisions; and
- how they think about and conceptualize the risk.

To accomplish this, those within the organization who interact directly with outside participants should be good listeners. They should not make facile prejudgments about what people think and know and which options they will prefer. They should be prepared for skepticism, antagonism, and hostility. They should respect the legitimacy of subjective, as distinct from coldly analytic, responses. They should not be surprised if people are more interested in matters of trust, credibility, and fairness than in the technical details of risk estimates and risk reduction options. They should not expect outside participants to know, or to necessarily accept, the legal or other practical boundaries that constrain the risk decision.

Risk managers should expect, and not resent (or appear to resent), skepticism about their motives in establishing more open procedures. They should understand that the fear of co-optation may impede trust, at least initially.

The job of interacting with outside participants should not be delegated to lower-level staff. Those with the power to make the decisions under discussion need to be directly involved in face-to-face dialogue, at least for the major issues, for this provides convincing evidence of the organization's sensitivity to the viewpoints of interested and affected groups.

In some cases it may be advisable to formalize the participation, for example, by forming a citizen advisory group. Such a move would signal an organizational commitment to continue to listen and to heed. Representative sample surveys also can help identify what people know and how they feel, what they think their choices are, and their responses to new information. Such surveys could constitute a valuable contribution to the openness of the overall process.

Early and Sustained Interaction

The best form of interaction is that which begins at an early stage and continues from then on. If outside groups are brought in very late, they are likely to be frustrated if the decisions the organization has already made are effectively off limits for discussion. Participating organizations have scarce resources and will resent being drawn into what they see as empty proceedings.

Open procedures are most successful when the host organization leaves itself ample room to adapt as discussions mature. For example, participants may want to see the underlying risk assessment done in a different way, so as to illuminate issues of particular salience (e.g., the risks or costs imposed on particular groups, alternative units of measurement). They may call for further data collection to address uncertainties that trouble them most. When participants are asked to contribute to the development of a risk message itself, they may want to explore different strategies for dissemination and additional target audiences. Where time and legal considerations permit, participants may productively help the risk manager to develop new or refine extant risk management options.

Once participation has begun, it is important to sustain it. Regular updates, newsletters, and briefings can reinforce the belief that the organization is responsive to input from participating groups.

There may be strong disincentives to early efforts at openness. For example, at early stages the organization's risk assessment may be unfinished; openness at this point could result in inconsistent information emanating from different sources within the host organization, which itself could undermine trust and credibility. We do not wish to deny that such complications exist; however, such considerations should not be permitted to automatically preclude early participation unless they clearly outweigh its considerable advantages.

The Empowerment Problem

Openness is not the same thing as empowerment. Risk managers should anticipate some confusion concerning the objectives of participation. In the past some host organizations have seen participation as a means to a narrow end—the development of better risk messages—while outside participants may believe that they have been given a full vote in making the risk management decision (e.g., the choice among regulatory options, the decision to issue a public health announcement, the decision to locate a new corporate facility) or in changing the decision process itself. Some participants may feel that discussing risk messages without addressing the risk management decision itself is beside the point. There is admittedly a fine line between being responsive to outside concerns and relinquishing responsibility to make risk management decisions. It is the risk manager's responsibility to be as clear as possible at the outset about where the line is drawn for a particular case. (This does not mean, however, that the risk manager should expect assent on this point, and ambiguity is likely to remain, but it is a matter that is better explicitly discussed than left below the surface.)

Outside participants need to understand that, because of statutes and electoral responsibility, an organization cannot, and should not, share its responsibility for risk management decisions. Federal agencies, for example, are not commonly able to delegate authority and still remain within their legal authority and thus accountable to the electorate (through executive or legislative oversight) for their regulatory actions. In the past vested interests have been suspected (often by groups that are absent because they cannot afford the costs of participation) of abusing open procedures to thwart or delay decisions, rather than to improve them. The host organization, for its part, should not expect participating groups to relinquish their right to raise objections later on, using litigation or other means, simply because they have been consulted in advance.

Safeguarding Balance and Accuracy in Risk Messages— Preventing Real and Perceived Distortion

For many risk messages, credibility depends on the audience's belief that the message is reasonably objective; there is broad skepticism about organizations shading the truth to suit their ends.

Because bias, like beauty, is often in the eye of the beholder, it may be very difficult for those who oversee the preparation of risk messages to ascertain, by examining the messages themselves, whether they will mislead audiences or be perceived as distorted. Procedural safeguards may be much more effective.

To help ensure that risk messages are not distorted and do not appear as distorted, those who manage the generation of risk assessments and risk messages should:

- hold the preparers of messages accountable for detecting and reducing distortion;
- consider review by recognized independent experts of the underlying assessment and, when feasible, the message;
- when feasible, subject draft messages to outside preview to determine if audiences detect any overlooked distortions; and
- prepare and release a "white paper" on the risk assessment and risk reduction assessment for comment.

Accountability

Distortion can enter at two stages: in the preparation of the expert analyses that form the basis of a risk message and in the composition of the message itself. Risk managers should actively encourage those who prepare messages and the expert analysts within the organization to supply materials that are as free of distortion as possible. Risk managers should sensitize employees to particular types of biases and perceived biases that it is particularly concerned about and see that the experts are aware of subtle causes of perceived bias.

Experts commonly must synthesize risk information that is fraught with uncertainty, for which many choices among competing quantitative and qualitative assumptions and methodologies must be exercised. To cite but one of the many assumptions that will be found in a particular case, for example, the assessment may be based on "worst case" or on "best estimate" calculations. There is a constant danger that such choices will be unduly influenced by three types of bias: (1) the expert's personal value judgments about what the risk management outcome should be; (2) the expert's belief of where the organization's self-interest lies; and (3) "expert bias," which sometimes leads experts to exaggerate the certainty and precision of their assessments. Unfortunately, one cannot assume that experts are significantly more self-conscious about the subtle distinction between value judgment and scientific consensus in complex analyses than nonexperts are; this means that the risk manager needs to be actively involved in preventing distortion in the way risk assessments and risk reduction assessments are performed and presented.

Some of the assumptions inherent in a risk assessment have to do with uncertainties in the underlying science: for example, choosing among available scientifically supportable theories about extrapolating to humans from animal data on the carcinogenicity of various doses of tested substances. These assumptions unavoidably interject a subjective element into the risk assessment, by reflecting the assessor's judgment about which extrapolation method is most likely to be confirmed by future research.

Other assumptions will reflect the values of the scientists performing the assessment. They may have chosen to use "conservative" estimates in various portions of their work. They may have summarized risks to different groups of exposed people in a way that ignores who those people are (rather than, say, giving extra weight to risks to children).

Risk analysts and risk managers also may make special assumptions about how to weight diverse risks. For example, risks that involve horrific outcomes (e.g., cancer deaths as compared to death by cardiovascular disease) and catastrophic outcomes (e.g., release of lethal chemicals in populated areas, as occurred in Bhopal, India) are sometimes given extra weight in making risk management decisions.

Such assumptions may be widely accepted value choices. They may be just what the public prefers experts to do when confronting uncertainty. However, they need routinely to be made explicit if audiences are to interpret the resulting risk messages appropriately. Moreover, because they reflect the interjection of values into assessment, they need to be cited as a matter of openness and public accountability.

More commonly feared by skeptical recipients than expert bias is the possibility of (intentional or unintentional) distortion to fit ideological precepts. Government organizations are particularly susceptible to suspicions of distortion born of ideological bias. Congress, the press, and advocacy groups frequently charge that agency positions subordinate science to the current administration's ideology (e.g., a preservationist tendency in the Carter administration and a laissez-faire one in the Reagan administration). One should not conclude that such influences are flatly inappropriate in public risk management decisions; we elect presidents and legislators based on their expressed values and platforms and then hold them politically accountable for the value judgments they make when they are in office. Thus, for example, different administrations may properly seek different balances between health risks and economic benefits.

Similarly, how agency decision makers value usually unmeasurable quantities (e.g., nonmarket goods) in reaching regulatory decisions appropriately depends on philosophy or ideology. However, when the ideology overrides science or blinds the decision maker to established facts, the result is distortion.

Risk managers usually rely on information provided to them by staff scientists, engineers, and analysts. Frequently, this information is generated several levels below the manager and must pass through a series of intermediate managerial and policy reviewers. These reviews can filter out information or positions that are seen to contradict current policies, presenting a danger that the risk manager receives, perhaps unknowingly, distorted, incorrect, or inadequate information. Risk managers should establish an environment in which staff members believe themselves obligated to be honest and to come forward with their best information and analysis, even if it is not entirely welcome. Risk managers should not permit anyone to be penalized for arguing within the organization against the organization's or the administration's position, when the facts point elsewhere. They should remain constantly aware that failure to elicit the best technical information from within the organization can be extremely counterproductive to their credibility. Establishing this environment may be abetted by a formal procedure, such as that established by the Nuclear Regulatory Commission in 1980, known as Differing Professional Opinions (U.S. Nuclear Regulatory Commission, NRC Manual Chapter NRC-4125, September 1980, amended July 1985), and by periodic attention to ensure effectiveness (NUREG-1290, "Differing Professional Opinions: 1987 Special Review Panel," U.S. Nuclear Regulatory Commission, November 1987).

Independent Review

To help ensure that choices made in performing the risk assessment do not introduce errors or analytic assumptions that conflict with areas of current scientific consensus, organizations should routinely subject the underlying assessments, and when feasible the ensuing risk messages themselves, to independent peer review. This review can help managers satisfy themselves that uncertainties are adequately characterized and that scientific disagreements are understood.

Peer review should be as independent of the communicating organization as possible and should be conducted by a group whose collective expertise blankets the scientific areas that are germane to the risk message. The Science Advisory Board of the U.S. Environmental Protection Agency (EPA) and various assessment panels convened by the National Research Council are effective examples.

Message Preview

When possible, drafts of risk messages and the information on which they are based should be made available to selected outside individuals for their preview and comment.

Previews by partisans is a proven method of identifying intentional and unintentional slants in risk messages: if value judgments have inappropriately intruded to produce distortion, groups that hold contrary values are certain to proclaim the misstep. In cases where early participation has been possible, the participants themselves can perform the preview.

Outside previews will also help reveal where agreement exists among diverse groups. Such coordination can help reduce the incidence of needlessly competing or conflicting messages from groups that are in basic accord.

There are many instances in which partisan preview is not advisable, particularly when it would appear that the organization is unfairly giving advance information on major policy changes to some groups and not to others. (Note: Previews of messages by the general public have also proven effective. "Focus groups" have increasingly been used, less to detect bias and inaccuracy than to judge whether the intended message is actually understood. Although more expensive and time consuming, representative sample surveys can be used to provide a more accurate picture of the likely response of the intended audience.)

Written Document

The assessment of potential bias, as well as the search for technical errors, is greatly enhanced when written supporting documents are available. When time and resources permit, the communicating organization should synthesize the scientific information base into a formal "white paper" that can be generally released. This document should summarize relevant quantitative and qualitative scientific information, the attendant uncertainty about the risk and about risk reduction alternatives, and the assumptions employed. Federal agencies could release such a document as—or in conjunction with—the preamble to a formal notice of proposed rule making, as has been done for major regulations by the Food and Drug Administration and the EPA. Such a document can facilitate the elicitation of reactions from review by independent experts, partisan groups, and even the lay public. It can also foster understanding of the risk issue in different parts of the organization itself.

Fostering Competence—Making Risk Communication Smarter

Risk communication has only recently come into focus as a concept, and in many organizations it is still subsumed under other functions, such as risk assessment or public affairs. More attention should be paid to risk communication as a distinct undertaking. Successful efforts in risk communication require a blend of technical and communications proficiency in the risk organization. Excluding technical experts can lead to false or incomplete messages or the appearance or reality of the willful manipulation of facts. Excluding those with public affairs functions provides a danger of insensitivity to the capacities, interests, and needs of the audience.

Risk managers need to use procedures that attain a balance between two distinct types of expertise: the risk subject matter (e.g., carcinogenic risk, occupational safety) and risk communication. Organizations that communicate about risk should take steps to ensure that the preparation of risk messages becomes a deliberate, specialized undertaking, taking care that in the process they do not sacrifice scientific quality. Such steps include:

- deliberately considering the makeup of the intended audience and demonstrating how the choice of media and message reflects an understanding of the audience and its concerns;
- attracting appropriate communications specialists and training technical staff in communications;
- requiring systematic assurance that substantive risk experts within the organization have a voice in producing accurate assessments and the derivative risk messages;
- establishing a thoughtful program of evaluating the past performance of risk communication efforts; and
- ensuring that their organizations improve their understanding of the roles of intermediaries, particularly media reporters and editors, including an understanding of the factors that make a risk story newsworthy, of the practical time and space constraints, and of the limited technical background of most media personnel.

Assessment of Audience

As noted above, a source organization should, before it initiates risk communication, set realistic goals; it should make a deliberate effort to formulate its communication objectives, identify intended audiences, consider alternative communication strategies, and assess the likely usefulness of a message to the audience.

Risk communication cannot be considered an informal add-on to the technical assessment effort. Effective risk communication involves specialized knowledge of, and when feasible interaction with, the intended target audience, an understanding of media practices, and an appreciation of the role of other intermediaries in relaying and translating messages. Those who assess risks and risk control options within an organization are not usually experienced in these areas.

As soon as the organization's risk communication objective is established, analysis of and interaction with the target community, or its representatives, should commence; deliberate audience research is important. Ideally, an audience profile should be compiled that describes the nature of the members of the audience and gives some idea of whom they trust, what they believe, and what concerns and worries motivate their actions; focus groups, surveys using representative sample techniques, and demographic studies may be helpful in compiling the profile. Available time and resources do not always permit the compilation of a detailed profile, but the risk manager should realize that risk communication will suffer to the extent that the audience is mischaracterized. The results of such audience research should be made public in a timely manner. Failure to do so may undermine the apparent openness of the organization.

Specific knowledge of the intended target groups permits intelligent segmentation of the audience, another key to effective communication. A uniform message will have varying effects on different individuals. Audience segmentation is useful both to customize the message and in the choice of the most effective communication channels. Risk communicators need to be aware that individuals may prefer to use different channels for different aspects of their decision-making process. Some channels, for example, may be best for conveying general knowledge but less reliable for affecting whether individuals believe a risk is or is not something to be worried about.

Risk managers should expect those who prepare risk messages to construct a communication plan that clearly links the choice of channel and customized messages to an understanding of audience segments and that links the definition of audience segments to an objective understanding of the audience population.

Two caveats should be noted. First, while an explicit analysis of the audience is important, it should not be expected to supplant all other considerations in planning for the risk message. Other important organizational goals (legal constraints, consistency of current and past policy, support of current enforcement efforts) must be factored in. The risk manager's difficult job is to attain a reasonable balance among these competing organizational objectives. Second, but no less important, the risk manager must be concerned with the outside appearance of the explicit communication planning effort. Observers may rightly or wrongly perceive a deliberate effort to understand, segment, and reach the audience as inherently manipulative and invasive.

Specialized Talent

Risk communication requires specialized knowledge and talent. It may be difficult to adequately reeducate technical or other existing staff to coordinate the message preparation effort. Preparing and helping implement the explicit risk communication plan described in the previous section require special expertise. Specialized knowledge in such subjects as demographic techniques, the psychology of risk perception, and how the media work, combined with the rare knack for writing clearly about complicated technical issues, is needed.

Recruiting staff with such capabilities—or retraining existing staff—amounts to putting the task of risk communication on a professional level in the organization in order to achieve better-informed risk communication decisions. However, skeptics inside and outside the organization may see it as importing dubious strategies and techniques from marketing and advertising into the heretofore scientific domain of risk assessment. Vigilance must be applied (open procedures can be of great value here) to ensure that such techniques do not become manipulative or deceptive in fact or appearance.

Scientific/Technical Accuracy and Completeness

Upgrading the staff that coordinates the preparation of risk messages to a professional level does not mean that substantive experts within the communicating organization can be shunted away from the process. They must remain involved in order to ensure that factual errors are not introduced.

Technical inaccuracy and incompleteness in message content can easily be used by knowledgeable advocates of alternative positions. For example, national advocacy groups, including consumer and environmentalist groups, use competent scientific/technical professionals in presenting positions and in countering government and corporate press releases. Once a message has been shown to be inaccurate or misleading, organizational credibility is lost for that message and for succeeding messages on quite different topics.

A technically flawed risk message may reflect poor risk communication *within* the communicating organization. To prevent this, risk managers should require that senior technical staff have an opportunity to evaluate the quantitative and qualitative accuracy of risk messages and that any exceptions are clearly reported.

Evaluation and Feedback

Even when communications professionals help design and guide the risk communication effort, doubt will remain about whether and how the intended audience will apprehend the message.

Source organizations should routinely conduct retrospective evaluations of their communication efforts and of particular messages. At this stage there appears to have been remarkably little formal evaluation by organizations that communicate about risks. Evaluation, if coupled to a feedback mechanism, is a necessary step in ensuring improvement in the competence of an organization's risk communication program. Organizations that disseminate risk messages should institute formal programs that assess their experience. Evaluations should address both questions of content and questions of process, as described in this report. That is, the effectiveness of messages should be examined—along with a sense of how different channels and intermediaries affected transmission—but attention also needs to be devoted to the organization's performance with its procedures for setting realistic goals, involving interested and affected parties, attaining balance, and creating internal expertise in risk communication.

Role of Intermediaries

Most risk messages will pass through one or more organizations or individuals before reaching the final recipients. Sometimes the only way of ensuring that a message reaches the people for whom it is intended is to rely on intermediaries. In any case an organization initiating risk communication should identify the intermediaries who will handle its messages, assess their needs and constraints, and adjust to those conditions if possible.

Journalists look first for clear statements about events and issues at conflict. They operate under strict deadlines and compete for allocations of space or time. Providing journalists with written copy will reduce, but not eliminate, the chances of being misquoted. Regular contact with journalists, including after stories have appeared, will generally improve the basis for later exchanges.

Community organizations and prominent individuals can be effective intermediaries for risk messages. But health departments, public libraries, professional associations, voluntary organizations, and similar groups all have their own aims and purposes. Discovering which organizations or individuals would be appropriate in a particular situation and developing the working relationship that is necessary for constructive interaction with such intermediaries can require considerable time and effort. It can, however, make the difference in reaching the intended recipients.

Some Notes on Handling Risk Communication in Crisis Conditions

Many risk situations require that risk messages be delivered immediately: examples include emergency conditions, challenges to an organization's positions before the organization is prepared to respond, and intense and contentious public controversy. In that atmosphere the deliberate procedures recommended above (e.g., outside reviews and analysis of the audience) may well be impractical.

The process for risk communication in crisis conditions requires special care. Risk managers should ensure that:

• where there is a foreseeable potential for emergency, advance plans for communication are drafted. These plans should be drafted jointly with the intended audiences (e.g., local communities near a chemical plant, paramedics, and fire departments). Such plans should be prepared in the context of concrete events and scenarios, should provide specific information that is relevant to people's risk-averting actions, and should specify actions that may be taken in case of a disaster or emergency; and

• there is provision for coordinating information among the various authorities that might be involved and, to the extent feasible, a single place where the public and the media can obtain authoritative and current information.

THE CONTENT OF RISK MESSAGES

The preceding section is addressed to risk managers, who have overall responsibility within their organization for assessing risks, making risk management decisions, and managing risk communications. This section is addressed to those within the communicating organization who are responsible for preparing formal risk messages.

In general, we find that practical advice on the content of risk messages depends heavily on the particular situation; for example, a public health advisory message on AIDS and an EPA announcement on the regulation of the use of a pesticide for certain crops may have quite different purposes, audiences, urgency, and visibility. We concentrate here on four generic matters—audience awareness, uncertainty, comparative risk, and completeness—that have been the source of difficulty in the past over a broad range of risk communication efforts.

Relating the Message to the Audiences' Perspectives

Risk messages are often based on the information in special analyses prepared for internal organizational purposes (e.g., to assess whether a particular risk exists or what risk management option to choose). That information often reflects the prior knowledge, perspectives, and language of risk experts and risk managers. It may not be sufficient for effective risk messages.

Risk messages should closely reflect the perspective, technical capacity, and concerns of the target audiences. A message should:

- **emphasize information relevant to any practical actions that individuals can take;**
- **be couched in clear and plain language;**
- **respect the audience and its concerns; and**
- **seek strictly to inform the recipient, unless conditions clearly warrant the use of influencing techniques.**

Personal Relevance

Consideration of the specific decisions that recipients face provides the surest basis for determining what risk information to emphasize in a risk communication. Such decisions might be whether and how to change personal behavior to respond to a reported health risk, whether to use or avoid a product that is being regulated, how to vote on a local siting issue, and whether to follow a particular risk issue further.

Much of the information available to those who prepare formal risk messages has been assembled in risk assessments prepared in a context of risk management decision making. The basic question in such assessments is, "What should the organization do (if anything) to reduce risk to the population?" Estimates of total exposures, total risk reduction costs, and other aggregate data—often written by experts whose immediate objective, understandably, is to make them scientifically defensible in the eyes of other experts—often predominate. The central question answered in a risk message should be "What should the recipient know to improve the choice among personal options (including the consequences of doing nothing)?" Data and analyses that risk experts have not emphasized may be needed. In the terms of decision theory, a risk message should contain information to which those decisions are "sensitive"—the facts that are most central to the choice at hand. This criterion should determine the kinds of information included and the detail and precision with which it is presented. For some decisions the critical information is the magnitude of the risk involved; for others it is the processes by which risks are created and controlled.

Risk information should be expressed in terms of risk to a representative individual, not only as a general population estimate. If there are highly exposed or particularly sensitive subgroups, such groups should be identified in a way that individuals can understand if they have reason for concern. Practical advice on such matters as danger signals of exposure, available remedies, sources of help, and so on should be included.

Selecting information relevant to individual choice is particularly important for risk messages—health warnings are prime examples—that are intended for an audience that is not already motivated to listen. The existence of such risks may mean little if it is not made clear what practical measures an exposed individual might use to avoid or reduce them.

Clarity

The risk message should be understandable to the target audience. When there is doubt about the ability of the audience to absorb technical material, little is lost by assuming that the audience has little technical training. Carefully chosen, vivid, concrete images and the use of personalized examples can help a lay audience to understand and can often even ensure understanding among those who are more familiar with the subject risk. Message designers should try to avoid using images or terms ("morbidity" is one example) that, while seemingly familiar to laypersons, have different or more precise technical meanings for experts.

Special care is needed in depicting statistical concepts and probabilities. Few people can meaningfully distinguish among small probabilities and may have no way of determining if such an assessment as "1-in-10,000 lifetime risk" is worth worrying about.

In a long message with extensive technical detail or quantitative complexity, the key portions—conclusions, summary, and recommended actions—should be written in lay terms.

The pursuit of clarity is likely to be enhanced by experimentation, post hoc evaluation, and the pretesting of messages with laypersons, all of which have been discussed above.

In our view clarity is a necessary but not a sufficient condition for improved risk messages. This is because while such features as plain language and vivid examples can enhance understanding, they can also, if misused, potently enhance *mis*understanding. One pitfall is that of equating clarity with brevity. The message preparer's goal should not be to gloss over the complexity and uncertainty of a risk but to reflect those qualities in plain language. Those who prepare risk messages should expect that their attempts at brevity will provoke protest among those who fear it will lead to greater misunderstanding. Where the nature of the chosen communication channel requires a short message, as with mass media announcements, this of course poses an unavoidably difficult dilemma and one in which the procedural measures recommended above (e.g., openness and message pretesting) may be vital.

Respect for the Audience and Its Concerns

If a message appears insensitive to an audience's actual concerns, there is a real chance that the audience will be alienated. The message should not disparage people's subjective reactions as inferior to expert assessments. If members of the audience hold beliefs that the source organization sees as false, it is better for the message to address these beliefs than to omit them as irrelevant. If people are advocating specific options for reducing risk, the message should address them, even if the source organization's analysis shows the options to be infeasible technically or legally. The impact of a message is, of course, determined as much by style and general demeanor as literal message content.

In all cases, but particularly in face-to-face delivery of messages, care should be taken to show compassion and to avoid distant, antiseptically statistical treatment of illnesses, injuries, and death. For example, if a person is gravely concerned about a particular hazard, a message that dismisses the risk as trivially small will surely come across as coldly patronizing.

The best way to summarize this general point is to observe that legitimacy is inherently reciprocal in nature; only if a source acknowledges the legitimacy of the audience's felt concerns will it have a chance to be seen as legitimate itself.

Use of "Influence Strategies"

Those who prepare risk messages, and particularly those in government organizations, need to be circumspect about using "influence strategies" in their risk messages to influence recipients' beliefs or actions, and they should expect their audiences to suspect attempts to influence even when the intent is simply to inform.

Americans are usually most comfortable with risk messages that, in Jefferson's words, "inform their discretion," but that do not attempt to advise them how to act in response. Some would draw a line for risk messages at the function of describing the risks and other outcomes (e.g., costs) associated with alternative risk management options, claiming that to go further involves the application of value judgments that are beyond the proper reach of the message source. Others would point out that governments make such judgments commonly, as when they by law or regulation establish sanctions against certain private actions (e.g., polluting, littering, and not wearing seat belts). In addition, Americans want their public servants to be strongly committed to the pursuit of their agency's national mission, and such individuals understandably form strong views on what they see as the correct ways to respond to problems. They will want to express those strong views. Not understanding many of the concepts presented in this report, these public servants may see a dilemma with respect to their role as risk communicators.

In Chapter 4 we introduced a distinction between two risk message strategies, informing and influencing. Influence strategies comprise a range of techniques ranging from, at one end, messages that attempt to "persuade" through the selective use of factual information to outright deception at the other extreme. The user's intent is to convince a recipient to accept the source's opinions or prescribed actions. The audience response to a message that is seen as using influence strategies may be to count it as illegitimate. The audience may disregard the entire risk message as slanted toward a predetermined outcome. Credibility is a casualty.

When is an influence strategy appropriate? As we noted in the discussion in Chapter 4, Americans accept influence strategies in some settings. Dietary warnings by public health officials are an example of the generally accepted use of influence. There is some indication that reigning traditions vary according to the culture of the professional field of the risk assessors—traditions in dam safety, for example, may differ from those in toxicology with respect to practitioners' efforts to prescribe policy or personal choices. Risk managers in government agencies would be well served to know when the use of influence strategies is safely legitimate and when it is not.

In general, we would urge great care in the use of influence strategies by government agencies. We have identified three particular situations in which use of influence strategies by government may arouse resentment that could affect the credibility of a message and/or source:

1. When there is unresolved public controversy over the issue, particularly if there has been no public forum at which relevant voices have had their say. Whenever government attempts to influence citizens' beliefs and actions, it should be able to point to some legitimate public process—one that has given interested and affected parties a chance to express themselves—which concluded that using risk messages to influence behavior serves an important public purpose.

2. When the form of influence strategy is toward the more severe end of the spectrum of influence techniques (i.e., near deception).

3. When there is no evident threat of externalized effects—that is, when the risk is confined largely to the persons who themselves undertake the risky behavior, without endangering others.

When influence strategies are used, risk messages should attempt to distinguish the analytic function of describing risk from the prescriptive function of advising recipients about what to do. It is an error to imply that the technical analysis led irrefutably to the prescription. The organization that disseminates the message should make clear that, in balancing risks and other factors to arrive at its recommended action, it has made a policy judgment. It is, of course, politically accountable for such judgments.

Handling Uncertainty

Uncertainty is a central fact in the assessment of many contemporary risks. It is usually present both in risk assessment and in the assessment of risk management options. The way that risk messages treat this uncertainty can have a major influence on the effectiveness and credibility of a communication effort. A major difficulty is avoiding unnecessary confusion between scientific uncertainty on one hand and policy disagreement on the appropriate risk management approach on the other hand.

Risk messages and supporting materials should not minimize the existence of uncertainty. Data gaps and areas of significant disagreement among experts should be disclosed. Some indication of the level of confidence of estimates and the significance of scientific uncertainty should be conveyed.

There are dangers if existing uncertainty is widely perceived as either underplayed or exaggerated. Any attempt to minimize uncertainty may make it appear that the caveats expressed by experts are being ignored. Exaggerating uncertainties can have the effect of obscuring the scientific basis of a risk management decision (e.g., whether to regulate, whether to issue a health advisory to the public), leaving the audience with the impression that the decision has been arbitrary in nature.

One reason the effectiveness of risk messages is so sensitive to their treatment of uncertainty is that the handling of uncertainty is a central issue in many of today's risk controversies. Often one side in a controversy will emphasize the need to base important risk management decisions on sound science, rather than on mere conjecture. As often, the other view will emphasize that ordinary prudence—"better safe than sorry"—dictates that action can be taken before conclusive scientific proof comes in. A central dispute thus becomes, "How much proof is needed?" and the degree of extant scientific proof itself becomes a matter of close partisan scrutiny.

Those who prepare risk messages commonly must choose between presenting the full range of available estimates, presenting a restricted set, or offering a single estimate based on consensus among consulted experts. Choosing any of these methods has its dangers, and more complicated presentations of the range of uncertainty are often needed; what is appropriate depends on the data available for a particular case.

It is usually dangerous for messages to characterize the overall level of uncertainty quantitatively, as might be done by describing statistical confidence intervals. In most situations expert assessments have multiple sources of uncertainty, and statistical measures do not adequately represent the complexity of the analysis.

For many messages an extensive description of uncertainty obviously cannot be included in the text itself. However, it remains useful to have prepared an explicit account, even if for practical reasons it must be consigned to supplementary documents made available to recipients upon request.

In general, those preparing risk messages are best served if they have available to them a statement of the scientific conclusions of the assessment of a professional quality that might be used for materials intended for expert peer review, such as papers submitted to professional journals; this will help ensure that uncertainties and necessary qualifications are adequately conveyed from the experts to those who prepare messages.

A form of sensitivity analysis can be helpful. To gauge the significance of uncertainty and of differences among experts, it is frequently helpful to vary the different sets of expert estimates systematically and then to gauge the effects on the overall risk estimate. The estimate will be more sensitive to some choices of assumptions than to others. If the risk message uses one of the competing assumptions, the risk message should say so, disclose why it was chosen over others, and indicate what difference it makes to the assessed risk. It should be observed that this procedure is one for which the needs of risk communication may dictate how a risk assessment itself is done.

The general goal of this recommendation is to help audiences distinguish areas of scientific agreement amid what may appear as vast areas of policy disagreement. The advantage of careful delineation of existing scientific uncertainty is that it gives audiences a sense of the degree of scientific consensus and allows them to distinguish minor from major uncertainties.

Comparing Risks

One factor that inhibits public understanding of risk messages is that people often cannot easily relate the low—say, 1 in 10,000—risk probabilities presented to their everyday experience. They are thus often deprived of a sense of the personal meaning of the risk in question and so cannot arrive at a comfortable decision about whether to take actions to deal with the risk or whether to be concerned at all about the hazard. In theory, at least, this difficulty can be overcome by quantitative comparisons between risks between familiar and less familiar risks).

Risk comparisons can be helpful, but they should be presented with caution. Risk comparisons must be seen as one of several inputs to risk decisions, not as determinants of decisions. There are proven pitfalls when risks of diverse character are compared, especially when the intent of the comparison can be seen as that of minimizing a risk (by equating it to a seemingly trivial risk). More useful are comparisons of risks that help convey the magnitude of a particular risk estimate, that occur in the same decision context (e.g., risks from flying and driving to a given destination), and that have a similar outcome. Multiple comparisons may avoid some of the worst pitfalls. More work needs to be done to develop constructive and helpful forms of risk comparison.

In theory, at least, comparative information should be an attractive element of risk messages. We have advised that the best risk messages are those that inform the recipient's actual choices, and increasingly those choices are between courses of action (or inaction) that represent different risks. Risk comparisons ideally might help individuals steer a prudent course between risks of various sizes.

However, actual attempts to compare risks have engendered considerable controversy and distrust. One reason for this is the fear that comparisons will be used to influence and even mislead the lay public. Individuals are known, for example, to subjectively underestimate actual incidence rates for some fatal risks (e.g., those resulting from asthma and strokes) and to overestimate others (e.g., risks that are especially feared, like those resulting from tornadoes and botulism). Thus, comparing a risk to the likelihood of death by asthma would probably induce most people to similarly underestimate it.

Another difficulty is that alternatives often have more than one risk attribute, and different people emphasize different facets. For a particular choice, for example, one group might concentrate on the relative number of deaths associated with each alternative, and a second group may emphasize the way risks (or costs) are distributed among various groups within society. The choice of any single metric for comparison will thus ignore facts that some observers may value highly.

Some who have used comparative risk information seem to have done so on the assumption that recipients would, upon seeing that a particular risk is small, elect to stop worrying about it. Implicit was the notion of an "action threshold," or perhaps a "worry threshold," that would render subthreshold risks unfit for serious consideration. The notion has a debilitating flaw, and it is not surprising that risk comparisons that seemed to be used to trivialize certain risks met with objections. Personal and organizational risk management decisions are based on many factors, of which a risk estimate is only one. For example, even a trivial risk may be worth eliminating if the costs of elimination are negligible; to suggest that people should decide based on one factor—for example, expected mortality—alone is somewhat analogous to saying people should make purchases based solely on comparative pricing without considering the value of the product to them. In practice, risk comparison data can rarely be closely linked to specific decisions in the absence of other critical information about decision options.

In general, comparisons of "unlike" risks should be avoided, as they have often either confused message recipients or irritated them because they were seen as unfair or manipulative. Directly comparing voluntary (e.g., skiing) and involuntary (e.g., air pollutants) risks, or natural (e.g., earthquakes) and technological (e.g., food additives) risks, for example, is rarely a good idea. More generally, those who prepare risk messages should appreciate the weakness of risk comparisons as a means of placating people about risks that are calculated to be small.

When can comparisons be used in a risk message? Three situations suggest themselves:

1. To help message recipients comprehend probabilities. In isolation a term like "one chance in a million per year" may convey little. An analogy to lengths (1 inch to 16 miles) or volumes (1 drop to 16 gallons) may help some people; reference to other known one-to-a-million risks of the type under discussion (for lung cancer, that of smoking a certain number of cigarettes; for private transportation mortality, that of traveling 300 miles by car) may help others, if they have a grasp of the reference risk.

2. To directly compare alternative options. Personal and organizational decisions can be better informed if the risks of alternative actions are laid out in comparable terms. Comparing the risks of coffee and tea consumption, or the risks of air and automobile travel between two points, may improve one's ability to make informed choices (again, however, one would not expect a risk comparison to necessarily dominate in such choices). For a regulatory agency the health risk of a pesticide may be directly compared to that of its substitute if it were removed from commerce.

3. To gauge the relative importance of different causes of the same hazard. Discussions of public and private actions with respect to indoor radon may be improved, for example, by a comparison of radon with smoking and other known causes of lung cancer.

One interesting approach is the use of risk ladders, for which a range of probabilities is presented for a single class of risks. The discussion of Figure 5.1 shows the limitations of past use. If one is careful, however, the use of multiple comparisons helps counteract the possibility that people may severely misestimate a particular risk, even though it is familiar to them. It also reduces the danger of arousing the scientific disputes that can often arise when only two risk estimates are compared, one or both of which are subject to scientific debate.

Ensuring Completeness

If the information in a risk message is incomplete, the recipients may be unable to make well-informed decisions.

A complete information base contains five types of qualitative and/or quantitative information: (1) the nature of the risk, (2) the nature of the benefits that might be affected if risk were reduced, (3) the available alternatives, (4) uncertainty in knowledge about risks and benefits, and (5) management issues. There are major advantages in putting the information base into written form as an adjunct to the risk message.

Those who prepare risk messages should ensure that the messages are complete. A suggested risk information checklist of relevant topics for the design of a complete message, drawn from the description in Chapter 2, is summarized in Figure 7.1.

Two points are worth emphasis. First, a complete risk message, as we have defined it, includes information other than a risk assessment; it covers the characterization of current or possible efforts to

INFORMATION ABOUT THE NATURE OF RISKS
1. What are the hazards of concern?
2. What is the probability of exposure to each hazard?
3. What is the distribution of exposure?
4. What is the probability of each type of harm from a given exposure to each hazard?
5. What are the sensitivities of different populations to each hazard?
6. How do exposures interact with exposures to other hazards?
7. What are the qualities of the hazard?
8. What is the total population risk?

INFORMATION ABOUT THE NATURE OF BENEFITS
1. What are the benefits associated with the hazard?
2. What is the probability that the projected benefit will actually follow the activity in question?
3. What are the qualities of the benefits?
4. Who benefits and in what ways?
5. How many people benefit and how long do benefits last?
6. Which groups get a disproportionate share of the benefits?
7. What is the total benefit?

INFORMATION ON ALTERNATIVES
1. What are the alternatives to the hazard in question?
2. What is the effectiveness of each alternative?
3. What are the risks and benefits of alternative actions and of not acting?
4. What are the costs and benefits of each alternative and how are they distributed?

UNCERTAINTIES IN KNOWLEDGE ABOUT RISKS
1. What are the weaknesses of available data?
2. What are the assumptions on which estimates are based?
3. How sensitive are the estimates to changes in assumptions?
4. How sensitive is the decision to changes in the estimates?
5. What other risk and risk control assessments have been made and why are they different from those now being offered?

INFORMATION ON MANAGEMENT
1. Who is responsible for the decision?
2. What issues have legal importance?
3. What constrains the decision?
4. What resources are available?

FIGURE 7.1   Risk message checklist.

reduce risk. Some topics include the cost of control, who pays, how effective the approach is, and whether the control implies additional risks of its own. Uncertainty in the analysis of risk control measures should be included. The message should also contain pertinent information about how any risk management decision has been or will be made.

Second, the checklist used for preparing a complete risk message should be used to ensure that the underlying analysis itself is complete; that is, concern for risk communication should influence the conduct of risk assessment and risk control assessment. If the information base developed in the analytic process is incomplete, the risk message will be deficient.

There are advantages to compiling and keeping the information base in written form. In at least some cases, for example, it will prove useful to compile a "white paper" of factual information on the subject risk. As described in the section above on management of the process, a written record provides a useful management tool for risk communication; if the underlying information is in written form, it can be examined (and perhaps improved) by others inside and outside the organization, helping to prevent surprises when the risk message is disseminated. Such a document also can provide a useful single source for diverse messages, enhancing consistency and accuracy. When feasible, this document should be made available as an adjunct to the formal risk message.

Whether or not the information base is compiled in written form, risk communicators should treat it—and be seen as treating it—as work in progress that is continually subject to improvement. Discussions and debates that surround a risk message often raise new questions, and new data can arise from research and other sources.

A CONSUMER'S GUIDE TO RISK AND
RISK COMMUNICATION

A major theme of this report is that risk communication should be understood to be a two-way interchange between source organizations and those, including the public and its representatives, who are the intended recipients of risk messages. In the previous pages we have directed many recommendations about the process and content of risk communications efforts to source organizations, specifically government agencies and large corporations.

If risk communication is a two-way enterprise, both sides have rights and responsibilities that must be understood if the process is to work well. The following recommendation is directed at improving the recipient's ability to participate meaningfully in risk management and risk communication. It is based on the conclusion that, at this stage, nonexpert participants have different understandings of the nature of risk and how it is managed. It is also based on the conclusion that the risk communication process would benefit if the interested public were better able to ask intelligent, probing questions of those in government, industry, and elsewhere who prepare risk messages for their consumption. As source organizations become more accomplished at risk communication, we expect that there will be more opportunities for two-way interactions. We believe there needs to be a national locus for improving the public's ability to participate.

Major government and private organizations, including environmental and consumer groups, that sustain risk communication efforts should jointly fund the development of a Consumer's Guide to Risk and Risk Communication. The purposes of this guide would be to articulate key terms, concepts, and trade-offs in risk communication and risk management for the lay audience, to make audiences better able to discern misleading and incomplete information, and to facilitate the needed general participation in risk issues.

Such a guide should:

• involve support from, but not control by, the federal government and other sources of risk messages;

• be under the editorial control of a group that is clearly oriented toward the recipients of risk messages, and under administrative management by an organization that is known for its independence and familiarity with lay perspectives, and that can undertake the needed outreach and public information effort; and

• cover subjects such as those suggested below—e.g., the nature of risk communication, the concepts of zero risk and comparative risk, and evaluating risk messages—and others designated by project participants.

We believe that the development of such a guide would have several advantages. It would help orient the interested public—and the leaders of organized groups—and prevent some of the misunderstanding that has occurred in the past. It would provide nonexpert participants with tools and concepts to enhance their participation, including sections about how to identify incomplete, imbalanced, or misleading messages. The process of writing it would advance national discussion about areas of current controversy among players in an often adversarial process of making risk management decisions. The guide would also articulate the basis for public skepticism that sometimes causes consternation among those responsible for risk management and the design of risk messages.

Project Support

It is important that major risk communicators—federal agencies, large corporations—support the project. We would expect that the project would require about 1 year to complete and that it would require a full-time staff of two or three persons. Allowance should be made for wide distribution of gratis copies of the final document. Provision should be made to update the guide 3 to 5 years after it is published; updating will help ensure that there is a national focal point for continuing interactions among the groups that, together, can bring about long-term improvement in risk communication.

Project Management

Editorial control of the guide should be exerted by a steering group in which the views and concerns of the lay recipients of risk messages are paramount. It should not be difficult to identify individuals who reflect an appropriately broad range of lay perspectives. The steering group should also include a minority of other relevant perspectives (e.g., risk managers, scientists and other experts, media representatives, and advocacy groups).

The project requires a stable but independent administrative home. For practical reasons it would be most suitably placed under the aegis of an existing organization in order to permit an efficient start-up and a reliable dissemination/outreach phase. The administrative home should be one that is credible to all sides involved in risk management issues and one that has demonstrable relevant experience. The League of Women Voters and the National Safety Council are two of several organizations that meet these criteria.

An integral part of the project should be the design of a dissemination effort that, among other possibilities, makes use of compatible existing efforts at public outreach involving aspects of risk by professional (e.g., American Bar Association, American Medical Association, American Chemical Society) and other groups.

Content of the Guide

We offer a brief topic list as representative of subjects to be covered in a consumer's guide (see Figure 7.2). In addition to coverage of these points—and other subjects raised during the guide project itself—the guide might contain a directory of information resources on risk topics for the lay public and groups that represent it.

WHAT IS RISK?
Key Terminology and Concepts
  Hazard, exposure, probability, sensitivity, individual risk,
  population risk, distribution of risk, unattainability of zero risk
Qualitative Attributes
  Voluntariness, catastrophic potential, dreadedness, lethality,
  controllability, familiarity, latency

WHAT DOES RISK ASSESSMENT CONTRIBUTE?
Quantification
  Quality, completeness, uncertainty, confidence
Scientific and Policy Inferences
  Assumptions, assessment of benefits, risk management choices

WHAT IS THE ROLE OF THE RISK COMMUNICATION PROCESS?
Setting
  Public debate about decisions, informing or influencing personal action
Purpose
  Messages can inform, influence, or deceive
Interaction Among Participants
  Contending conclusions, justifications, credibility, and records

HOW CAN YOU FIND OUT WHAT YOU NEED TO KNOW?
Technical Content
  Demystifying jargon, comparing relevant risks, finding trusted interpreters
Independent Sources
  Information clearinghouses, academic or public service sources

HOW CAN YOU PARTICIPATE EFFECTIVELY?
Finding the Right Arena
  Identifying the responsible decision maker, getting on the agenda
Intervention
  Identifying points and times for intervention, marshalling support

HOW CAN YOU EVALUATE THE MESSAGES AND THE COMMUNICATORS?
Accuracy
  Factual base, track record, consistency, self–serving framing, use of
  influence techniques, misleading risk comparisons
Legitimacy
  Standing, access, review, due process justification
Interpreting Advocacy
  Comparing competing arguments, seeing where information has been omitted,
  questioning message sources

FIGURE 7.2 A consumer's guide to risk and risk communication.

RESEARCH NEEDS

As a result of our deliberations, we recommend the nine specific research topics listed below. Some stem directly from the problems identified in Chapter 6. Others are based on our review of available information and the substantial practical experience of committee members. Two criteria guided our selection of topics: (1) additional knowledge would lead to material improvement in risk communication practices and (2) creation of such knowledge is likely, given past results and current research methods. We have not set priorities among the topics.

Risk Comparison

If performed thoughtfully, risk comparison holds promise of making risk communication more relevant and meaningful to recipients. However, three issues need to be explored to prevent past shortcomings of the technique.

• Comparability. When are two risks "similar" enough in nature to be compared without misleading, confusing, or angering recipients? What are the crucial dimensions across which risks should not be compared?

• Apprehension of risk magnitudes. How do people apprehend the magnitudes of risks; in particular, how do they interpret very small probabilities, which often seem beyond most people's intuitive understanding? How do different ways of presenting risk magnitudes affect people's feeling for the size of risks?

• Validation. The use of risk comparisons is undermined if there is doubt about the validity of the data that are compared. Risk estimates used in risk comparisons must be validated in two ways: (1) as to the current scientific accuracy and the associated uncertainty or qualifications and (2) as to whether nonexperts are known to systematically underestimate or overestimate such estimates subjectively (which would make them inappropriate as "anchors" in risk comparisons).

Risk Characterization

We need better ways of presenting complex information about risk clearly and accurately and better understanding of the limitations of techniques for simplifying complex material. How do people respond to alternative ways of characterizing risks, including alternative treatments of uncertainty?

Role of Message Intermediaries

We need a better empirical base for understanding the role of intermediaries in carrying and translating risk messages. What channels (mass media, specialized media, advocacy groups, community organizations, local professionals and other opinion leaders, casual acquaintances) do people actually use? How do people validate and integrate messages from multiple sources in deciding what to do, or what to believe, about a particular risk? Case examinations and a review of research in allied fields (e.g., medical education) can help elucidate the direct and indirect flow of information from source to recipient.

Pertinency and Sufficiency of Risk Information

Risk communicators need to focus on the information that is most pertinent to recipients' needs; they are in danger of wasting the limited access they have to their audience if they are viewed as preoccupied with marginal issues. What types of information do people actually find pertinent in reaching personal decisions about risk? How does this compare with what the risk manager or decision analyst thinks *should* be pertinent? How and when do people determine that they do not need additional information in order to decide what they will do about a risk? What information appears necessary to trigger active personal concern about a risk?

Psychological Stress

Given the number and variety of known risks in modern life, what conditions are necessary to induce stress about a particular risk in persons and communities? Which of the messages that appeal to fear, or that advert to imminent danger, actually cause stress? If people are stressed about a particular risk, how is their apprehension of risk information affected?

Recipients' "Mental Models"

The information in risk messages is useful only if recipients can incorporate it into their prior thinking about the risk and its management. Only by better knowing how recipients conceptualize risks and their risk decisions can people create more effective messages. In particular:

• How do people think about the risk decisions that confront them? For example, what alternatives do they consider, and what consequences are they aware of?

• How do people think about the causal processes that create risks? For example, do they misconceive exposure processes, and how effective do control efforts seem, intuitively?

• How do people perceive the social/governmental processes involved in managing risks? For example, what do they believe regulatory agencies are empowered to do, and when are public interest advocates seen as credible?

Risk Literacy

How do people learn the "analytic" concepts and language they need to understand risk statements? Do they lack important concepts? What kinds of materials, including special curricular materials in science and mathematics education, might be effective?

Retrospective Cases

There is a dearth of case studies that focus directly on risk communication. In particular, retrospective case materials should be prepared that:

• Examine risk communication *processes*, including such topics as the role of experts and others in message preparation, whether and how outside groups were involved in risk management and risk communication decisions, and the role of intermediaries in message transmission.

• Analyze the *responses of recipients* and how the responses corresponded to the expectations of the source.

Contemporaneous Assessments of Risk Cases

Too seldom are there attempts to learn from ongoing cases of risk management. This is partly due to an understandable desire to concentrate resources on solving a risk problem, rather than calibrating it; nonetheless, real-time assessments can provide valuable knowledge for making general improvements in risk communication. This contemporaneous research should address such matters as how people react to different types of messages and channels; what their actual concerns, frustrations, and data needs are; and how effective alternative communication and message strategies are.

Appendixes

Appendix A
Background Information on
Committee Members and
Professional Staff

COMMITTEE MEMBERS

JOHN F. AHEARNE, *Chairman*, is vice president of Resources for the Future, Washington, D.C. A physicist specializing in systems and policy analysis in defense, energy, and resources, Dr. Ahearne served as a deputy assistant secretary of defense for systems analysis, deputy assistant secretary of defense for program analysis and evaluation, and principal deputy assistant secretary of defense for manpower and reserve affairs. Dr. Ahearne also served as systems analyst for the White House Energy Office (1977) and deputy assistant secretary of energy for resource applications (1978). He was a member of the Nuclear Regulatory Commission from 1978 to 1983 and was chairman from 1979 to 1981. Currently, he is chairman of the Department of Energy's Advisory Committee on Nuclear Facility Safety.

ERNESTA BALLARD is a private consultant on toxic substance management in Seattle, Washington. As a regional administrator for the Environmental Protection Agency from 1983 to 1986, she was responsible for implementation and enforcement of environmental programs in Alaska, Idaho, Oregon, and Washington. Ms. Ballard served as director of public services for Seattle (1976–1978) and budget director of the University of Washington (1974–1976). She is chairman of the Board of Trustees of University Hospital, University of Washington; a member of the Advisory Board of Albers School of

Business, Seattle University; and a member of the Board of Trustees of The Nature Conservancy.

RUTH FADEN is professor of health policy and management, the Johns Hopkins University, Baltimore, Maryland, where she directs the Program in Law, Ethics, and Health, and is senior research scholar, Kennedy Institute of Ethics, Georgetown University. Dr. Faden has done extensive research and writing in ethics and health policy and is coauthor of the book *A History and Theory of Informed Consent*. She has served as a consultant to, among others, the National Institute on Alcohol Abuse and Alcoholism, Office of Technology Assessment, President's Commission for the Study of Ethical Problems in Medicine and Biomedical and Behavioral Research, and Centers for Disease Control.

JAMES A. FAY is professor of mechanical engineering, Massachusetts Institute of Technology. Dr. Fay's areas of expertise include air pollution and energy. He has been a member of Maine's Natural Resources Council and the Massachusetts Energy Facility Siting Council, as well as chairman of the Boston Air Pollution Control Commission. Dr. Fay served on the National Research Council's Environmental Studies Board and the Committee on Radioactive Waste Management. He is a fellow of the American Institute of Aeronautics and Astronautics, the American Academy of Arts and Sciences, and the American Association for the Advancement of Science; a member of the American Society of Mechanical Engineers; and director of the Union of Concerned Scientists.

BARUCH FISCHHOFF is professor in the Department of Engineering and Public Policy and the Department of Social and Decision Sciences at Carnegie-Mellon University, Pittsburgh, Pennsylvania. Earlier, Dr. Fischhoff spent 11 years with Decision Research and Eugene Research Institute, Eugene, Oregon, working in the areas of judgment and decision making, human factors, and risk management. He has numerous publications in these fields, including the book *Acceptable Risk*. Dr. Fischhoff is on the editorial boards of *Policy Sciences, Cognitive Psychology, Journal of Personality and Social Psychology, Accident Analysis and Prevention, Social Behavior Organizational Behavior and Human Decision Processes*, and *International Journal of Forecasting*. He has served on the National Research Council's Committee on Priority Mechanisms for the National Toxicology Program, Panel on Survey Measurement of Subjective Phenomena, Committee on Human Factors, and Committee on Pilot Performance Modeling for a Computer Aided Design and Engineering Facility.

THOMAS P. GRUMBLY is president of Clean Sites, Inc., Alexandria, Virginia. Mr. Grumbly was executive assistant to the commissioner of the U.S. Food and Drug Administration (1977–1979); deputy administrator, Food Safety and Inspection Service, U.S. Department of Agriculture (1979–1981); and staff director, Subcommittee on Investigations and Oversight, House Committee on Science and Technology, U.S. House of Representatives (1981–1982). He spent three years as executive director of the Health Effects Institute. Mr. Grumbly has also served as a consultant to the U.S. Environmental Protection Agency in the area of risk assessment and has served on the National Research Council's Panel on Reform of the Federal Meat and Poultry System.

PETER BARTON HUTT is a partner with the law firm of Covington & Burling, Washington, D.C. His expertise is in administrative and regulatory law. He served as chief counsel to the Food and Drug Administration (1971–1975). He is a member of the Institute of Medicine and serves on the advisory boards of the Institute for Health Policy Analysis, Georgetown University; the Scripps Clinic and Research Foundation, La Jolla, California; and the Center for Study of Drug Development, Tufts University. Mr. Hutt has served on a number of National Institutes of Health, Institute of Medicine, and National Research Council committees and on the advisory panels on technology innovation and health, safety and environmental regulation, animal testing, biotechnology, and medical devices for Congress's Office of Technology Assessment. He coauthored the book *Food and Drug Law: Cases and Materials* and has written a number of book chapters and articles. He has also worked and written extensively in the area of drug and alcohol abuse. Mr. Hutt serves on the editorial boards of *Regulatory Toxicology and Pharmacology, Food, Drug and Cosmetic Law Journal,* and *Biotechnology Law Report.*

BRUCE KARRH is vice president, safety, health and environmental affairs, E. I. du Pont de Nemours & Co., Wilmington, Delaware. He is a fellow of the American Academy of Occupational Medicine, the American College of Preventive Medicine, and the American Occupational Medical Association. Dr. Karrh is chairman of the Board of Directors of the Chemical Industry Institute of Toxicology and of the American Industrial Health Council. He is also a member of the Board of Directors of Thomas Jefferson University and its clinical affairs committee. He is a diplomate of the American Board of Preventive Medicine, certified in occupational medicine.

D. WARNER NORTH is a principal with Decision Focus, Inc., Los Altos, California, specializing in Decision Analysis; he is also a consulting professor, Department of Engineering-Economic Systems at Stanford University, and associate director of Stanford's Center for Risk Analysis, Stanford, California. Over the last 20 years Dr. North has carried out applications of decision analysis and risk assessment to a variety of public policy issues. He has participated in six previous National Research Council studies on air quality and toxic chemicals, including the 1983 National Research Council's Committee on the Institutional Means for Assessment of Risks to Public Health. His recent work includes development of decision frameworks for risk management of coal combustion by-products and acid deposition. Dr. North is a member of the Scientific Advisory Panel to the Governor of California for the Safe Drinking Water and Toxic Enforcement Act of 1986 (Proposition 65). He has served on committees of the Science Advisory Board of the U.S. Environmental Protection Agency since 1979.

JOANN E. RODGERS is deputy director of public affairs and director of media relations, The Johns Hopkins Medical Institutions, Baltimore, Maryland. Ms. Rodgers worked as a newspaper journalist specializing in science writing for 20 years and continues as a freelance writer of books and magazine articles on science and medicine. She is a past president of the National Association of Science Writers and a vice president of the Council for the Advancement of Science Writing. She is the recipient of a number of science writing awards, including a Lasker Award, two American Heart Association awards, and the AMA Medical Journalism award. She is the author or coauthor of books on drugs and childrearing and hundreds of magazine articles. She teaches and lectures frequently on science communication.

MILTON RUSSELL is professor of economics and senior fellow, Waste Management Research and Education Institute, University of Tennessee, Knoxville, and senior economist, Oak Ridge National Laboratory, Oak Ridge, Tennessee. He served as assistant administrator for policy, planning and evaluation at the U.S. Environmental Protection Agency (1983–1987). Dr. Russell was senior fellow and director of the Center for Energy Policy Research at Resources for the Future and spent 2 years as a senior staff economist with President Ford's Council of Economic Advisors. He taught in Iowa and

Texas before joining the Economics Department at Southern Illinois University, which he subsequently led as chairman. Dr. Russell has coauthored some half-dozen books focusing on energy and resource economics, lectured widely, and authored over 70 articles and chapters in journals and texts.

ROBERT SANGEORGE is vice president for public affairs of the National Audubon Society, headquartered in New York City. He was a working journalist for 12 years (1972–1984), including 3 years as the national environment and energy correspondent for United Press International, based in Washington, D.C. He held several other assignments during 9 years of service with UPI, including supreme court correspondent and bureau chief in Cleveland, Ohio. He also worked as a reporter/producer for 3 years in public broadcasting. Prior to his present position with the National Audubon Society, he was the assistant to the president for public accountability of Clean Sites, Inc., Alexandria, Virginia. Mr. SanGeorge held a Kiplinger Foundation Fellowship at Ohio State University in 1975–1976.

HARVEY M. SAPOLSKY is professor of public policy and organization in the Political Science Department, Massachusetts Institute of Technology. Dr. Sapolsky specializes in bureaucratic politics and science and public policy. He has studied and written articles on risk, specifically concerning cigarette smoking, the fluoridation of water, and AIDS and the blood supply, and has recently edited the book *Consuming Fears: The Politics of Product Risks.* Dr. Sapolsky is a fellow of the American Association for the Advancement of Science and a member of the American Political Science Association and is on the editorial boards of the *Journal of Health Politics, Policy and Law* and *Inquiry.*

JURGEN SCHMANDT is professor, LBJ School of Public Affairs, University of Texas, and director, Center for Growth Studies, Houston Area Research Center. He has published books on nutrition policy, the acid rain dispute between Canada and the United States, and environmental and resource policies. He recently served on the Texas Science and Technology Council. While serving as a senior environmental fellow at the U.S. Environmental Protection Agency, he worked on the development of a strategy for the control of toxic substances in the environment. From 1965 to 1970, Dr. Schmandt was associate director of Harvard University's Program on Technology and Society. At the Organization for Economic Cooperation and

Development in Paris he directed the review series on science policy in member countries.

MICHAEL SCHUDSON is professor, Department of Communication and Department of Sociology, and chair, Department of Communication, University of California, San Diego. Dr. Schudson's areas of expertise are the media and advertising. He is the author of the books *Discovering the News: A Social History of American Newspapers, Advertising, the Uneasy Persuasion*, and *Reading the News*, as well as many articles on the media. Dr. Schudson also serves as corresponding editor for *Theory and Society* and is a member of the editorial board of *Critical Studies in Mass Communication*.

PERCY H. TANNENBAUM is professor of public policy and director of the Survey Research Center, University of California, Berkeley. His research specialties include communication behavior, attitude change and measurement, mass media functions and effects, telecommunications policy, and social research methodology. His recent books include *Tuned-on TV/Turned-off Votes: Policy Options for Election Projections* and *Flies in the Policy Ointment: Perspectives in the California Medfly Crisis*. Dr. Tannenbaum is a fellow of the American Association for the Advancement of Science and the American Psychological Association and was a resident fellow at the Center for Advanced Study in the Behavioral Sciences (Stanford) and the Institute for Advanced Study, Berlin.

DETLOF von WINTERFELDT is director, Risk Communication Laboratory, and professor, Department of Systems Science, Institute of Safety and Systems Management, University of Southern California. His specialties are decision theory and risk analysis. Dr. von Winterfeldt coauthored *Risk Communication: A Review of the Literature* and *Decision Analysis and Behavioral Research*. He is also a member of the Institute of Management Science and the Society for Risk Analysis, an associate member of Operations Research Society of America, an associate editor for *Operations Research*, and a member of the editorial board of *Risk Analysis* and *Risk Abstracts*.

CHRIS WHIPPLE is technical manager, Risk and Health Science Department, Environment Division, at the Electric Power Research Institute, Palo Alto, California. Dr. Whipple's expertise is in the areas of analysis and management of technological risks. He serves on the National Research Council's Board on Radioactive Waste Management and has served on National Research Council committees on the Health and Ecological Impacts of Synfuel Industries and on Nuclear Safety Research. He also chaired the International Atomic Energy Agency's Coordinated Research Program on Risk Criteria for the Nuclear Fuel Cycle. He served on the Advisory Committee to the National Science Foundation Project on Risk Assessment and has been a contractor with the U.S. Environmental Protection Agency's Office of Air Quality Planning and Standards. Dr. Whipple is a member and past president of the Society for Risk Analysis and a member of the American Association for the Advancement of Science.

SUSAN WILTSHIRE is senior associate at JK Associates, Hamilton, Massachusetts, a consulting firm specializing in public policy formulation and citizen involvement in technical decisions. Ms. Wiltshire is particularly involved in issues of radioactive waste management. She is a member of the National Research Council's Board on Radioactive Waste Management and served on the Board's Panel on Uranium Mill Tailings. She was a member of the Program Review Committee for the National Low-Level Waste Management Program and the Environmental/Institutional Review Group for the Office of Crystalline Repository Development. Ms. Wiltshire is former president of the League of Women Voters of Massachusetts, served as a member of the National League of Women Voters Nuclear Energy Education Project Advisory Committee, and coauthored the 1985 revision of the League of Women Voters' *A Nuclear Waste Primer*. She is currently chairman of the elected Board of Selectmen of the Town of Hamilton and vice-chairman of Northeast Health Systems, Inc., and of Beverly Hospital, Beverly, Massachusetts.

PROFESSIONAL STAFF

ROB COPPOCK is senior program officer with the Commission on Physical Sciences, Mathematics, and Resources of the National Research Council. Dr. Coppock was a staff scientist at the Science Center Berlin in West Germany before joining the commission. He has conducted research in the area of risk and regulation for several years and is the author of *Regulating Chemicals in Japan, West Germany, France, the United Kingdom, and the European Community: A Comparative Examination* and *Social Constraints on Technological Progress*. He edited, with others, *Technological Risk: Its Perception and Handling in the European Community*. From 1981 to 1987 he was on the editorial board of the journal *Risk Analysis*.

LAWRENCE E. McCRAY is executive director of the Committee on Science, Engineering, and Public Policy of the Academies and the Institute of Medicine. He was associate executive director of the Commission on Physical Sciences, Mathematics, and Resources of the National Research Council until August 1, 1988. Dr. McCray has held positions with the U.S. Environmental Protection Agency, the U.S. Regulatory Council, and the Office of Management and Budget. He was project director for the 1983 National Research Council Study on Risk Assessment in the Federal Government and a 1985 National Research Council Study on the Atmospheric Effects of Nuclear Explosions. Dr. McCray won the Schattschneider Award of the American Political Science Association for the best dissertation in American government and politics in 1973.

PAUL C. STERN is senior staff officer with the Commission on Behavioral and Social Sciences and Education of the National Research Council. He also is study director of the National Research Council's Committee on Contributions of Behavioral and Social Science to the Prevention of Nuclear War. He previously served as study director of the Committee on Behavioral and Social Aspects of Energy Consumption and Production at the National Research Council and as research associate at Yale University's Institution for Social and Policy Studies. Dr. Stern's current research is on the formation of social attitudes about environmental policy. He is coeditor of *Energy Use: The Human Dimension* and coauthor of the chapter "Managing Scarce Environmental Resources" in the *Handbook of Environmental Psychology*. He also chairs the Environmental Problems Committee of the Division of Population and Environmental Psychology, American Psychological Association.

Appendix B
Bibliography

Abelson, R., and A. Levi. 1985. Decision making and decision theory. In Handbook of Social Psychology, 3d ed., G. Lindzey and E. Aronson, eds. New York: Random House.

Ackerman, B. A., and W. T. Hassler. 1977. Clean Coal, Dirty Air. New Haven, Conn.: Yale University Press.

Ahearne, J. 1987. Nuclear power after Chernobyl. Science 236(4802):673–679.

Allen, F. W. 1987. Towards a holistic appreciation of risk: The challenge for communicators and policymakers. Science, Technology, and Human Values 12(3&4):138–143.

Ames, B. N., L. S. Gold, and R. Magaw. 1987a. Letter. Science 237(4821):1399–1400.

Ames, B. N., R. Magaw, and L. S. Gold. 1987b. Ranking possible carcinogenic hazards. Science 236(4799):271-280.

Appelbaum, P. S., C. W. Lidz, and A. Meisel. 1987. Informed Consent: Legal Theory and Clinical Practice. New York: Oxford University Press.

Aries, P. 1974. Western Attitudes Toward Death. Baltimore, Md.: Johns Hopkins University Press.

Arrow, K. J. 1982. Risk perception in psychology and economics. Economic Inquiry 20(1):1–9.

Ashford, N. A., C. W. Ryan, and C. C. Caldart. 1983. A hard look at federal regulation of formaldehyde: A departure from reasoned decisionmaking. Harvard Environmental Law Review 7:297–370.

Atkinson, S. E., T. D. Crocker, and R. G. Murdock. 1985. Have priors in aggregate air pollution epidemiology dictated posteriors? Journal of Urban Economics 17:319–334.

Baird, B. N. R. 1986. Tolerance for environmental health risks: The influence of knowledge, benefits, voluntariness, and environmental attitudes. Risk Analysis 6(4):425–435.

Bandura, A. 1978. The self system in reciprocal determinism. American Psychologist (April):344–358.

Barkdoll, G. L. 1983. Involving constituents in agency priority setting: A case study. Evaluation and Program Planning 6:31–37.

Barles, B., and J. Kotas. 1987. Pesticides and the nation's ground water. EPA Journal 13(4):42–43.

Bartlett, J. 1980. Familiar Quotations, 15th ed. Boston: Little, Brown.

Bean, M. C. 1987. Tools for environmental professionals involved in risk communication at hazardous waste facilities undergoing siting, permitting, or remediation. Paper presented at the 80th Annual Meeting of the Association Dedicated to Air Pollution Control and Hazardous Waste Management, New York, June 21–26, 1987.

Bean, M. C. 1988. Speaking of risk. Civil Engineering (February):59–61.

Bean, M. C., and M. K. Null. 1988. A Workshop for Citizens on Risk Assessment. EPA pilot project. Reston, Va.: CH2M Hill.

Benson, H., L. Gordon, C. Mitchell, and V. Place. 1977. Patient education and intrauterine contraception: A study of two package inserts. American Journal of Public Health 67(5):446–449.

Berreth, D. 1987. Presentation to National Research Council Committee on Risk Perception and Communication meeting, Washington, D.C., November 17, 1987.

Beyth-Marom, R. 1982. How probable is probable? Journal of Forecasting 1:257–269.

Breslow, L., S. Brown, and J. van Ryzin. 1986. Letter. Science 234(4779):923.

Brickman, R., S. Jasanoff, and T. Ilgen. 1985. Controlling Chemicals: The Politics of Regulation in Europe and the United States. Ithaca, N.Y.: Cornell University Press.

Broome, T. H. 1986. The slippery ethics of engineering. The Washington Post, December 28:D3a.

Brown, G. 1987. The outlook for a new pesticides law. EPA Journal 13(4):35–36.

Budescu, D. V., and T. S. Wallsten. 1987. Subjective estimation of precise and vague uncertainties. In Judgmental Forecasting, G. Wright and P. Ayton, eds. New York: John Wiley & Sons.

Burnham, D. 1976. Energy Agency Data Termed Misleading. New York Times, September 30:A13.

Campbell, G., and K. O. Ott. 1979. Statistical evaluation of major human errors during the development of new technological systems. Nuclear Science and Engineering 71:267–279.

Campt, D. 1987. Daminozide: A case study of a pesticide controversy. EPA Journal 13(4):32–34.

Carpenter, S. L., and W. J. D. Kennedy. 1988. Managing Public Disputes. San Francisco, Calif.: Jossey-Bass.

Carson, R. 1962. Silent Spring. Boston, Mass.: Houghton Mifflin.

Chaiken, S. 1980. Heuristic versus systematic information processing and the use of source versus message cues in persuasion. Journal of Personality and Social Psychology 39(5):752–766.

Chemical and Engineering News. 1985. Union Carbide: New accidents revive safety issue. (August 19):4.

Chemical Education for Public Understanding Project. 1986. Risk Module Teachers Guide (draft). University of California at Berkeley, November 18, 1986.

Chemical Manufacturers Association. 1986. CAER Progress Report. Washington, D.C.: Chemical Manufacturers Association.

Cialdini, R. B. 1984. Influence: How and Why People Agree to Things. New York: Morrow.

Clark, W. C. 1980. Witches, floods, and wonder drugs: Historical perspectives on risk management. In Societal Risk Assessment: How Safe Is Safe Enough?, R. C. Schwing and W. A. Albers, Jr., eds. New York: Plenum Press.

Cohen, B. L. 1987. Reducing the hazards of nuclear power: Insanity in action. Physics and Society 16(3):2–4.

Cohen, J. 1962. The statistical power of abnormal-social psychological research: A review. Journal of Abnormal and Social Psychology 65(3):145–153.

Cohen, J. L. 1985. Strategy or identity: New theoretical paradigms and contemporary social movements. Social Research 52:663–716.

Coppock, R. 1987. Risk perception and communication. Working paper for the National Research Council Committee on Risk Perception and Communication meeting, Washington, D.C., May 26–27, 1987.

Council on Environmental Quality. 1985. Report of an Expert Meeting on Research Needs and Opportunities at Federally-Supervised Hazardous Waste Site Clean-Ups, Council on Environmental Quality, Washington, D.C.: October 28–30, 1985.

Covello, V. T. 1984. Actual and perceived risk: A review of the literature. In Technological Risk Assessment, P. F. Ricci, L. A. Sagan, and C. G. Whipple, eds. The Hague: Martinus Nijhoff.

Covello, V. T., and M. Abernathy. 1984. Risk analysis and technological hazards: A policy-related bibliography. In Technological Risk Assessment, P. F. Ricci, L. A. Sagan, and C. G. Whipple, eds. The Hague: Martinus Nijhoff.

Covello, V. T., and F. Allen. 1988. Seven Cardinal Rules of Risk Communication. Washington, D.C.: U.S. Environmental Protection Agency, Office of Policy Analysis.

Covello, V. T., D. von Winterfeldt, and P. Slovic. 1986. Risk communication: A review of the literature. Risk Abstracts 3(4):171–182.

Covello, V. T., L. B. Lave, A. Moghissi, and V. R. R. Uppuluri, eds. 1987a. Uncertainty in Risk Assessment, Risk Management, and Decision Making. New York: Plenum Press.

Covello, V. T., P. Slovic, and D. von Winterfeldt. 1987b. Risk Communication: A Review of the Literature. Washington, D.C.: National Science Foundation.

Covello, V. T., P. M. Sandman, and P. Slovic. 1988. Risk Communication, Risk Statistics, and Risk Comparisons: A Manual for Plant Managers. Washington, D.C.: Chemical Manufacturers Association.

Crouch, E. A. C., and R. Wilson. 1982. Risk/Benefit Analysis. Cambridge, Mass.: Ballinger.

Cvetkovich, G., C. Vlek, and T. C. Earle. In press. Designing public hazard communication programs about large-scale technologies. In Social Decision Methodologies for Technological Projects, C. Vlek and G. Cvetkovich, eds. Amsterdam: North-Holland.

Davies, J. C., V. T. Covello, and F. W. Allen, eds. 1987. Risk Communication. Washington, D.C.: The Conservation Foundation.

Davis, D. 1987. Presentation to National Research Council Committee on Risk Perception and Communication meeting, Washington, D.C., November 17, 1987.

Davis, D. L., A. D. Lilienfeld, A. Gittelsohn, and M. E. Scheckenbach. 1986. Increasing trends in some cancers in older Americans: Fact or artifact? Toxicology and Industrial Health 2(1):127-144.

Deisler, P. F., Jr. 1988. The risk management-risk assessment interface. Environment, Science, and Technology 22(1):15-19.

Department of Health and Human Services. 1986. Determining Risks to Health: Federal Policy and Practice. Dover, Mass.: Auburn House Publishing.

Derbaix, C. 1983. Perceived risk and risk relievers—an empirical investigation. Journal of Economic Psychology 3:19-38.

Dezern, J. N. 1988. Risk assessment and ASTM: Reasons for and against ASTM's involvement. ASTM Standardization News 88(February):52-55.

Diamond, S. 1985. Carbide blames a faulty design for toxic leak; effects can be serious, company memo says. New York Times, August 13:A1, B8.

Dickson, D. 1984. The New Politics of Science. New York: Pantheon.

Dickson, R. B. 1987. Risk assessment and the law: Evolving criteria by which carcinogenicity risk assessments are evaluated in the legal community. In Uncertainty in Risk Assessment, Risk Management, and Decision Making, V. T. Covello, L. B. Lave, A. Moghissi, and V. R. R. Uppuluri, eds. New York: Plenum Press.

Dietz, T. M., and R. W. Rycroft. 1987. The Risk Professionals. New York: Russell Sage Foundation.

Dietz, T., P. C. Stern, and R. W. Rycroft. 1989. Definitions of conflict and the legitimation of resources: The case of environmental risk. Sociological Forum 4(1):47-70.

Dirkin, G. R. 1983. Cognitive tunneling: Use of visual information under stress. Perceptual and Motor Skills 56:191-198.

Douglas, M. 1985. Risk Acceptability According to the Social Sciences. Social Research Perspectives. New York: Russell Sage Foundation.

Douglas, M., and A. Wildavsky. 1982. Risk and Culture. Berkeley: University of California Press.

Dreman, D. 1979. Contrarian investment strategy. New York: Random House.

Dunlap, R. E. 1987. Public opinion and the environment in the Reagan era. Environment 29(July/August):6-11, 32-37.

Dunlap, R. E., J. K. Grieneeks, and M. Rokeack. 1983. Human values and pro-environmental behavior. In Energy and Material Resources: Attitudes, Values, and Public Policy, W. D. Conn, ed. AAAS Selected Symposium 75. Boulder, Colo.: Westview.

Dunwoody, S., M. Friestad, and M. A. Shapiro. 1987. Conveying risk information in the mass media. Paper presented to the Mass Communication Division of the International Communication Association, May 1987.

Eagly, A. H., and S. Chaiken. 1985. Psychological theories of persuasion. In Advances in Experimental Social Psychology, L. Berkowitz, ed. New York: Academic Press.

Economist, The. 1987. Making company disasters less disastrous (January 31):55-56.

Edwards, W., and D. von Winterfeldt. 1986. Public disputes about risky technologies: Stakeholders and arenas. In Risk Evaluation and Management, V. T. Covello, J. Menkes, and J. Mumpower, eds. New York: Plenum Press.

Elliott, M. L. P. 1987. The effect of differing assessments of risk in hazardous waste facility siting negotiations. Paper presented at the Workshop on Negotiating Hazardous Waste Facility Siting and Permitting Agreements, The Conservation Foundation, Washington, D.C., June 15, 1987.

Environmental Protection Agency. 1987a. Answering questions about pesticides: An interview with John A. Moore. EPA Journal 13(4):4–8.

Environmental Protection Agency. 1987b. A consumer's guide to safer pesticide use. EPA Journal 13(4):9–31.

Environmental Protection Agency, Science Advisory Board. 1988. Letter of March 9.

Faden, R. 1987. Ethical issues in government sponsored public health campaigns. Health Education Quarterly 14(1):27–37.

Faden, R. R., and T. L. Beauchamp. 1986. A History and Theory of Informed Consent. New York: Oxford University Press.

Ferguson, E. S. 1987. Risk and the American engineering profession: The ASME boiler code and American industrial safety standards. In The Social and Cultural Construction of Risk, B. B. Johnson and V. T. Covello, eds. Dordrecht, Holland: D. Reidel.

File, S. E., and A. Jew. 1973. Syntax and the recall of instructions in a realistic situation. British Journal of Psychology 64:65–70.

Fischhoff, B. 1982. Debiasing. In Judgment Under Uncertainty: Heuristics and Biases, D. Kahneman, P. Slovic, and A. Tversky, eds. New York: Cambridge University Press.

Fischhoff, B. 1983. "Acceptable risk": The case of nuclear power. Journal of Policy Analysis and Management 2(4):559–575.

Fischhoff, B. 1984. Setting standards: A systematic approach to managing public health and safety risks. Management Science 30(7):823–843.

Fischhoff, B. 1985a. Managing risk perceptions. Issues in Science and Technology 2(1):83–96.

Fischhoff, B. 1985b. Protocols for environmental reporting: What to ask the experts. The Journalist (Winter):11–15.

Fischhoff, B. 1985c. Risk analysis demystified. NCAP News (Winter):30–33.

Fischhoff, B. 1987. Treating the public with risk communications: A public health perspective. Science, Technology, and Human Values 12(3&4):13–19.

Fischhoff, B. 1988. Judgment and decision making. In The Psychology of Human Thought, R. J. Sternberg and E. E. Smith, eds. New York: Cambridge University Press.

Fischhoff, B., and L. A. Cox, Jr. 1985. Conceptual framework for regulatory benefits assessment. In Benefits Assessment: The State of the Art, J. D. Bentkover, V. T. Covello, and J. Mumpower, eds. Dordrecht, Holland: D. Reidel.

Fischhoff, B., and D. MacGregor. 1983. Judged lethality: How much people seem to know depends upon how they are asked. Risk Analysis 3:229–236.

Fischhoff, B., and O. Svenson. 1987. Perceived risks of radionuclides: Understanding public understanding. In Radionuclides in the Food Chain, G. Schmidt, ed. New York: Praeger.

Fischhoff, B., P. Slovic, and S. Lichtenstein. 1977. Knowing with certainty: The appropriateness of extreme confidence. Journal of Experimental Psychology: Human Perception and Performance 20:159–183.

Fischhoff, B., P. Slovic, S. Lichtenstein, S. Read, and B. Combs. 1978. How safe is safe enough? A psychometric study of attitudes towards technological risks and benefits. Policy Sciences 9:127–152.

Fischhoff, B., P. Slovic, and S. Lichtenstein. 1980. Knowing what you want: Measuring labile values. In Cognitive Processes in Choice and Decision Behavior, T. Wallsten, ed. Hillsdale, N.J.: Erlbaum.

Fischhoff, B., S. Lichtenstein, P. Slovic, S. L. Derby, and R. L. Keeney. 1981a. Acceptable Risk. New York: Cambridge University Press.

Fischhoff, B., P. Slovic, and S. Lichtenstein. 1981b. Lay foibles and expert fables in judgments about risk. In Progress in Resource Management and Environmental Planning, T. O'Riordan and R. K. Turner, eds. New York: John Wiley & Sons.

Fischhoff, B., P. Slovic, and S. Lichtenstein. 1983. "The public" vs. "the experts": Perceived vs. actual disagreements about risks of nuclear power. In Analysis of Actual vs. Perceived Risks, V. Covello, G. Flamm, J. Rodericks, and R. Tardiff, eds. New York: Plenum Press.

Fischhoff, B., S. R. Watson, and C. Hope. 1984. Defining risk. Policy Sciences 17:123-129.

Fischhoff, B., O. Svenson, and P. Slovic. 1986. Active responses to environmental hazards: Perceptions and decision making. In Handbook of Environmental Psychology, D. Stokols and I. Altman, eds. New York: John Wiley & Sons.

Fischhoff, B., L. Furby, and R. Gregory. 1987. Evaluating voluntary risks of injury. Accident Analysis and Prevention 19(1):51-62.

Fishburn, P. C. 1982. Foundations of risk measurement: II. Effects of gains on risk. Journal of Mathematical Psychology 25:226–242.

Fisher, A. 1987. Radon projects at EPA. Presented to National Research Council Committee on Risk Perception and Communication meeting, Washington, D.C., July 23, 1987.

Fiske, S., and S. Taylor. 1984. Social Cognition. Reading, Mass.: Addison-Wesley.

Fitchen, J. M., J. S. Heath, and J. Fessenden-Raden. 1987. Risk perception in community context: A case study. In The Social and Cultural Construction of Risk, B. B. Johnson and V. T. Covello, eds. Dordrecht, Holland: D. Reidel.

Folkman, S. 1984. Personal control and stress and coping processes: A theoretical analysis. Journal of Personality and Social Psychology 46(4):839–852.

Freudenburg, W., and E. Rosa, eds. 1984. Public Reaction to Nuclear Power: Are There Critical Masses? Boulder, Colo.: Westview.

Friedman, B., D. Lockwood, L. Snowden, and D. Zeidler. 1986. Mass media and disaster: Annotated bibliography. Miscellaneous Report No. 36. Newark: University of Delaware, Disaster Research Center.

Friedman, S., C. M. Gorney, and B. P. Egolf. 1987. Reporting on radiation: A content analysis of Chernobyl coverage. Journal of Communication 37(3):58–79.

Gale, R. P. 1987. Calculating risk: Radiation and Chernobyl. Journal of Communication 37(3):68–79.

Geller, E. S. 1983. Development of industry-based strategies for motivating seat belt usage. Final report for contract no. DTRS5681-C-0032. Washington, D.C.: U.S. Department of Transportation.

Gibbs, L. 1982. Love Canal: My Story. New York: Grove Press.

Gough, M., R. Hart, B. W. Karrh, A. Koestner, R. Neal, D. Parkinson, F. Perera, K. E. Powell, and H. S. Rosenkranz. 1984. Report on the consensus workshop on formaldehyde. Environmental Health Perspectives 58:323–381.

Gould, L. C., G. T. Gardner, D. R. DeLuca, A. R. Tiemann, L. W. Doob, and J. A. J. Stolwijk. 1988. Perceptions of Technological Risks and Benefits. New York: Russell Sage Foundation.

Gray, J. 1981. Three case studies of organized responses to chemical disasters. Miscellaneous Report No. 29. Newark: University of Delaware, Disaster Research Center.

Greenberg, D. S., and M. Taylor (illustrations). 1984. What is an acceptable risk? National Wildlife (August/September):29–32.

Greenwald, A. G., and C. Leavitt. 1984. Audience involvement in advertising: Four levels. Journal of Consumer Research 11:581–592.

Greenwood, T. 1984. Knowledge and Discretion in Government Regulation. New York: Praeger.

Hadden, S. G., ed. 1984. Risk Analysis, Institutions, and Public Policy. Port Washington, N.Y.: Associated Faculty Press.

Hadden, S. G. 1986. Read the Label: Providing Information to Reduce Health and Safety Risks. Boulder, Colo.: Westview Press for the American Association for the Advancement of Science.

Hance, B. J., C. Chess, and P. M. Sandman. 1988. Improving Dialogue with Communities: A Risk Communication Manual for Government. Trenton: Division of Science and Research Risk Communication Unit, New Jersey Department of Environmental Protection.

Harness, R. L. 1987. Managing pesticides: An industry view. EPA Journal 13(4):40–41.

Hasher, L., and R. T. Zachs. 1984. Automatic and effortful processes in memory. Journal of Experimental Psychology: General 108:356–388.

Hays, S. P. 1987. Beauty, Health, and Permanence: Environmental Politics in the United States, 1955-1985. New York: Cambridge University Press.

Henrion, M., and B. Fischhoff. 1986. Assessing uncertainty in physical constants. American Journal of Physics 54(9):791–798.

Hershey, J. C., and P. J. H. Shoemaker. 1980. Risk taking and problem context in the domain of losses: An expected utility analysis. Journal of Risk and Insurance 47:111–132.

Hively, W. 1988. Nuclear power at risk. American Scientist 76(July–August): 341–343.

Hogarth, R. M. 1981. Beyond discrete biases: Functional and dysfunctional aspects of judgmental heuristics. Psychological Bulletin 90(2):197–217.

Hohenemser, C., R. W. Kates, and P. Slovic. 1983. The nature of technological hazard. Science 220(4595):378-384.

Holland, N. 1987. Presentation to National Research Council Committee on Risk Perception and Communication meeting, Washington, D.C., November 17, 1987.

Hovland, C. I., I. L. Janis, and H. H. Kelley. 1953. Communication and Persuasion: Psychological Studies of Opinion Change. New Haven, Conn.: Yale University Press.

Humber, J. M., and R. F. Almeder, eds. 1987. Quantitative Risk Assessment: Biomedical Ethics Reviews–1986. Clifton, N.J.: Humana Press.

Hutt, P. B. 1974. A regulator's viewpoint. In How Safe Is Safe? The Design of Policy on Drugs and Food Additives. Washington, D.C.: National Academy Press.

Hutt, P. B. 1982. Food and drug law: A strong and continuing tradition. Food Drug Cosmetic Law Journal 37:123–137.

Hutt, P. B., and P. B. Hutt II. 1984. A history of government regulation of adulteration and misbranding of food. Food Drug Cosmetic Law Journal 39:2–73.

Ikeda, S. 1986. Managing technological and environmental risks in Japan. Risk Analysis 6(4):389–401.

Inglehart, R. 1977. Values, objective needs, and subjective satisfaction among western publics. Comparative Political Studies 4:428–458.

Inhaber, H. 1981. The risk of producing energy. Proceedings of the Royal Society of London A376:121–131.

Inside EPA. 1982. Congressmen will press EPA on national dioxin policy. (November 19):1, 4.

Institute for Environmental Negotiation. 1984. Not in my backyard! Community Reaction to Locally Unwanted Land Use. Charlottesville: University of Virginia.

Isaacs, T. 1987. Presentation to National Research Council Committee on Risk Perception and Communication meeting, Washington, D.C., July 23, 1987.

Jaeger, J. 1988. Developing Policies for Responding to Climatic Change. World Climate Programme Impact Studies (WMO/TD-No. 225). World Meteorological Association and United Nations Environmental Programme. April.

Jasanoff, S. 1986. Risk management and political culture. New York: Russell Sage Foundation.

Jasanoff, S. 1987. EPA's regulation of Daminozide: Unscrambling the messages of risk. Science, Technology, and Human Values 12(3&4):116–124.

Jenkins, C. 1983. Resource mobilization theory and the study of social movements. Annual Review of Sociology 9:527–553.

Jerome, F. 1986. Check it out: Journalists communicating about risk. Technology in Society 8:287–290.

Johnson, B. B., and V. T. Covello, eds. 1987. The Social and Cultural Construction of Risk: Essays on Risk Selection and Perception. Dordrecht, Holland: D. Reidel.

Kahneman, D., and A. Tversky. 1972. Subjective probability: A judgment of representativeness. Cognitive Psychology 3:430–454.

Kahneman, D., and A. Tversky. 1979. Prospect theory: An analysis of decision under risk. Econometrica 47(2):263–291.

Kahneman, D., P. Slovic, and A. Tversky, eds. 1982. Judgments Under Uncertainty: Heuristics and Biases. New York: Cambridge University Press.

Kasperson, R. 1986. Six propositions on public participation and their relevance for risk communication. Risk Analysis 6(3):275–281.

Kaufman, D. G. 1988. Assessment of carcinogenicity: Generic issues and their application to diesel exhaust. In Air Pollution, the Automobile, and Public Health, A. Y. Watson, R. R. Bates, and D. Kennedy, eds. Washington, D.C.: National Academy Press.

Keeney, R., and D. von Winterfeldt. 1986. Improving risk communication. Risk Analysis 6(4):417–424.

Kerr, R. A. 1988. Indoor radon: The deadliest pollutant. Science 240(4852):606–608.

Kong, A., G.O. Barnett, F. Mosteller, and C. Youtz. 1986. How medical professionals evaluate expressions of probability. New England Journal of Medicine 315(12):740–744.

Koshland, D. E. 1987. Immortality and risk assessment. Science 236(4799):241.

Krimsky, S., and A. Plough. 1988. Environmental Hazards: Communicating Risks as a Social Process. Dover, Mass.: Auburn House.

Lave, L. B. 1987. Health and safety risk analyses: Information for better decisions. Science 236(4799):291–295.

Lawless, E. W. 1977. Technology and Social Shock. New Brunswick, N.J.: Rutgers University Press.

Levine, A. G. 1982. Love Canal: Science, Politics, and People. Lexington, Mass.: Lexington Books.

Lewis, H. W. 1980. The safety of fission reactors. Scientific American 342(3):33–45.

Lichtenstein, S., and B. Fischhoff. 1980. Training for calibration. Organizational Behavior and Human Performance 26:149–171.

Lichtenstein, S., P. Slovic, B. Fischhoff, M. Layman, and B. Combs. 1978. Judged frequency of lethal events. Journal of Experimental Psychology: Human Learning and Memory 4:551–578.

Lichtenstein, S., B. Fischhoff, and L. D. Phillips. 1982. Calibration of probabilities: The state of the art. In Judgments Under Uncertainty: Heuristics and Biases, D. Kahneman, P. Slovic, and A. Tversky, eds. New York: Cambridge University Press.

Lind, N., ed. 1988. Risk Communication: A Symposium. Waterloo: University of Waterloo.

Lipset, S. M., and W. Schneider. 1987. The Confidence Gap: Business, Labor, and Government in the Public Mind (revised edition). Baltimore, Md.: Johns Hopkins University Press.

Lowrance, W. 1976. Of acceptable risk. San Francisco: Freeman.

Lynn, F. M. 1986. The interplay of science and values in assessing and regulating environmental risks. Science, Technology, and Human Values 11(2):40–50.

Lynn, F. M. 1987. Citizen involvement in hazardous waste sites: Two North Carolina success stories. Environmental Impact Assessment Review 7:347–361.

Maccoby, N., and D. S. Solomon. 1981. Heart disease prevention: Community studies. In Public Communication Campaigns, R. E. Rice and W. J. Paisley, eds. Beverly Hills, Calif.: Sage.

Manning, W. W. 1986. Concern justified about Frostban. Salinas Californian. January 22, 1986.

Marshall, E. 1983a. House reviews EPA's record on pesticides. Science 219(4589):1200.

Marshall, E. 1983b. Hit list at EPA? Science 219(4590):1303.

Marshall, E. 1983c. EPA's troubles reach a crescendo. Science 219(4591):1402–1404.

Mazis, M. B., R. Staelin, H. Beales, and S. Salop. 1981. A framework for evaluating consumer information regulation. Journal of Marketing 45:11–21.

Mazur, A. 1981. The Dynamics of Technical Controversy. Washington, D.C.: Communications Press.

Mazur, A. 1987. Putting radon on the public's risk agenda. Science, Technology, and Human Values 12(3&4):86–93.

Mazur, A. 1988. Mass Media Effects on Public Opinion About Nuclear Power Plants. Unpublished manuscript. Syracuse University, Syracuse, New York.

McAlister, A. 1981. Antismoking campaigns: Progress in developing effective communications. In Public Communication Campaigns, R. E. Rice and W. J. Paisley, eds. Beverly Hills, Calif.: Sage.

McArthur, L. Z., D. Q. Crocker, and E. Folino. 1981. Individual differences in cue utilization on spatial tasks. Perceptual and Motor Skills 52:923–929.

McCormick, N. J. 1981. Reliability and Risk Analysis. New York: Academic Press.

McGuire, W. J. 1985. Attitudes and attitude change. In Handbook of Social Psychology, 3d ed., Vol. 2., G. Lindzey and E. Aronson, eds. New York: Random House.

McKean, K. 1985. Decisions, decisions. Discover (June):22–31.

Melnick, R. S. 1983. Regulation and the Courts: The Case of the Clean Air Act. Washington, D.C.: Brookings Institution.

Melnick, R. S. 1988. The politics of cost-benefit analysis. Paper prepared for National Academy of Sciences Conference on Valuing Health Risks, Costs, and Benefits in Environmental Decisions, Washington, D.C., June 23–24, 1987.

Metropolitan Life Insurance Company. 1987. New high expectation of life. Statistical Bulletin 68(3):8–14.

Milbrath, L. 1984. Environmentalists: Vanguard for a New Society. Albany: State University of New York Press.

Mitchell, R. C. 1980. Public opinion on environmental issues. In Environmental Quality: The Eleventh Annual Report of the Council on Environmental Quality. Washington, D.C.: U.S. Government Printing Office.

Morrison, D. 1987. A Tale of Two Toxicities. Presentation to the Annual Meeting of the American Sociological Association, Chicago, Ill., August 1987.

Moscovici, S. 1985. Social influence and conformity. In Handbook of Social Psychology, 3d ed., Vol. 2., G. Lindzey and E. Aronson, eds. New York: Random House.

Mott, L. 1987. Managing pesticides: An environmentalist view. EPA Journal 13(4):37–39.

Murphy, A. H., and B. G. Brown. 1983. Forecast terminology: Composition and interpretation of public weather forecasts. Bulletin of the American Meteorological Society 64:13–22.

Murphy, A. H., and R. L. Winkler. 1984. Probability of precipitation forecasts. Journal of the American Statistical Association 79:391–400.

Murphy, A. H., S. Lichtenstein, B. Fischhoff, and R. L. Winkler. 1980. Misinterpretations of precipitation probability forecasts. Bulletin of the American Meteorological Society 61:695–701.

Mydans, S. 1987. Specter of Chernobyl looms over Bangladesh. The New York Times, June 5:I9.

Naber, T. 1988. Nanograms are not the answer to 'Will I be hurt?' questions. Waste Age (March):44–48.

National Research Council. 1979. Disasters in the Mass Media: Proceedings of the Committee on Disasters and the Mass Media Workshop. Washington, D.C.: National Academy Press.

National Research Council. 1980. The Effect on Populations of Exposure to Low Levels of Ionizing Radiation. Washington, D.C.: National Academy Press.

National Research Council. 1982. Risk and Decision-Making: Perspectives and Research. Washington, D.C.: National Academy Press.

National Research Council. 1983a. Risk Assessment in the Federal Government: Managing the Process. Washington, D.C.: National Academy Press.

National Research Council. 1983b. Risk Assessment in the Federal Government: Managing the Process. Working papers for the Committee on the Institutional Means for Assessment of Risks to Public Health. Washington, D.C.: National Academy Press.

National Research Council. 1984. Toxicity Testing: Strategies to Determine Needs and Priorities. Washington, D.C.: National Academy Press.

National Research Council. 1986a. Confronting AIDS: Directions for Public Health, Health Care, and Research. Washington, D.C.: National Academy Press.

National Research Council. 1986b. Drinking Water and Health, Vol. 6. Washington, D.C.: National Academy Press.

National Research Council. 1988a. Complex Mixtures: Methods for In Vivo Toxicity Testing. Washington, D.C.: National Academy Press.

National Research Council, Committee on the Biological Effects of Ionizing Radiations (BEIR IV). 1988b. Appendix II: Cellular Radiobiology. In Health Risks of Radon and Other Internally Deposited Alpha-Emitters, BEIR IV. Washington, D.C.: National Academy Press.

Nelkin, D. 1979a. Science, technology, and political conflict: Analyzing the issues. In Controversy: Politics of Technical Decisions, D. Nelkin, ed. Beverly Hills, Calif.: Sage.

Nelkin, D., ed. 1979b. Controversy: Politics of Technical Decisions. Beverly Hills, Calif.: Sage.

Nelkin, D. 1987. Selling Science: How the Press Covers Science and Technology. New York: W. H. Freeman.

Nelkin, D., and M. S. Brown. 1984. Worker at Risk: Voices from the Workplace. Chicago, Ill.: University of Chicago.

Neutra, R. R. 1985. Epidemiology for and with a distrustful community. Environmental Health Perspectives 62:393–397.

Nisbett, R. E., and L. Ross. 1980. Human Inference: Strategies and Shortcomings of Social Judgment. Englewood Cliffs, N.J.: Prentice-Hall.

Nuclear Regulatory Commission. 1985. NRC Manual Chapter NRC-4125 (September 1980, amended July 1985). Washington, D.C.: Nuclear Regulatory Commission.

Nuclear Regulatory Commission. 1987. Differing Professional Opinions: 1987 Special Review Panel. NUREG-1290. Washington, D.C.: Nuclear Regulatory Commission.

O'Brien, D. M., and D. A. Marchand. 1982. The Politics of Technology Assessment. Lexington, Mass.: Lexington Books.

Office of Communication of the United Church of Christ v. *Federal Communications Commission*, U.S. Court of Appeals District of Columbia Circuit, 359 F.2d 994 (1966).

Office of Technology Assessment. 1981. Assessment of Technologies of Determining Cancer Risks from the Environment. Washington, D.C.: U.S. Government Printing Office.

Okrent, D. 1980. An Approach to Quantitative Safety Goals for Nuclear Power Plants. NUREG-0739. Washington, D.C.: Nuclear Regulatory Commission.

Okrent, D. 1981. Industrial risks. Proceedings of the Royal Society of London A376:133–149.

Okrent, D. 1987. The safety goals of the U.S. nuclear regulatory commission. Science 236(4799):296–300.

O'Leary, M. K., W. D. Coplin, H. B. Shapiro, and D. Dean. 1974. The quest for relevance. International Studies Quarterly 18:211–237.

Otway, H. 1987. Experts, risk communication, and democracy. Risk Analysis 7(2):125–129.

Otway, H. J., and D. von Winterfeldt. 1982. Beyond acceptable risk: On the social acceptability of technologies. Policy Sciences 14:247–256.

Page, T. 1981. A framework for unreasonable risk in the Toxic Substances Control Act. In Carcinogenic Risk Assessment, R. Nicholson, ed. New York: New York Academy of Sciences.

Paté, M. E. 1983. Acceptable decision processes and acceptable risks in public sector regulations. IEEE Transactions on Systems, Man, and Cybernetics SMC-13(2):113–124.

Patterson, J. T. 1987. The Dread Disease. Cambridge, Mass.: Harvard University Press.

Peterson, C. R., and L. R. Beach. 1969. Man as an intuitive statistician. Psychological Bulletin 69:29–46.

Petty, R. E., J. T. Cacioppo, C. Sedikides, and A. J. Strathman. 1988. Affect and persuasion. American Behavioral Scientist 31(3):355–371.

Plough, A., and S. Krimsky. 1987. The emergence of risk communication studies: Social and political context. Science, Technology, and Human Values 12(3&4):4–10.

Pochin, E. E. 1980. The need to estimate risks. Physics in Medicine and Biology 25(1):1–12.

Pochin, E. E. 1981. Quantification of risk in medical procedures. Proceedings of the Royal Society of London A376:87–101.

Pochin, E. E. 1982. Risk and medical ethics. Journal of Medical Ethics 8:180–184.

Pollatsek, A., and A. Tversky. 1970. A theory of risk. Journal of Mathematical Psychology 7:540–553.

Poulton, E. C. 1968. The new psychophysics: Six models for magnitude estimation. Psychological Bulletin 69:1–19.

Poulton, E. C. 1982. Biases in quantitative judgments. Applied Ergonomics 13:31–42.

Quarantelli, E. L., D. C. Hutchinson, and B. D. Phillips. 1983. Evacuation Behavior: Case Study of the Taft, Louisiana Chemical Tank Explosion Incident. Miscellaneous Report No. 34. Newark: University of Delaware, Disaster Research Center.

Raiffa, H. 1968. Decision Analysis. Reading, Mass.: Addison-Wesley.

Rayner, S. 1984. Disagreeing about risk: The institutional cultures of risk management and planning for future generations. In Risk Analysis, Institutions, and Public Policy, S. G. Hadden, ed. Port Washington, N.Y.: Associated Faculty Press.

Rayner, S. 1986. Management of radiation hazards in hospitals: Plural rationalities in a single institution. Social Studies of Science 16:573–591.

Rayner, S. 1987a. Learning from the blind men and the elephant, or seeing things whole in risk management. In Uncertainty in Risk Assessment, Risk Management, and Decision Making, V. T. Covello, L. B. Lave, A. Moghissi, and V. R. R. Uppuluri, eds. New York: Plenum Press.

Rayner, S. 1987b. Risk and relativism in science for policy. In The Social and Cultural Construction of Risk, B. B. Johnson and V. T. Covello, eds. Dordrecht, Holland: D. Reidel.

Rayner, S., and R. Cantor. 1987. How fair is safe enough?: The cultural approach to societal technology choice. Risk Analysis 7(1):3–9.

Regens, J. L., and J. A. Donnan. 1986. Uncertainty and information integration in acidic deposition policymaking. The Environmental Professional 8(4):342–350.

Reich, R. B. 1985. Public administration and public deliberation: An interpretive essay. The Yale Law Journal 94:1617–1641.

Reilly, W. K. 1987. Foreword. In Risk Communication, J. C. Davies, V. T. Covello, and F. W. Allen, eds. Washington, D.C.: The Conservation Foundation.

Report of the Public's Right to Information Task Force. 1979. Washington, D.C.: U.S. Government Printing Office.

Reporting from the Russell Sage Foundation. 1987. Living with risk. Report No. 10, May 1987:1, 8, 11.

Ricci, P. F., L. A. Sagan, and C. G. Whipple, eds. 1984. Technological Risk Assessment. The Hague: Martinus Nijhoff.

Rice, R. E., and W. J. Paisley. 1985. Public Communication Campaigns. Beverly Hills, Calif.: Sage.

Roberts, D. F., and N. Maccoby. 1985. Effects of mass communication. In Handbook of Social Psychology, 3d ed., Vol. 2, G. Lindzey and E. Aronson, eds. New York: Random House.

Rodricks, J. V., S. M. Brett, and G. C. Wrenn. Undated. Significant Risk Decisions in Federal Regulatory Agencies. Washington, D.C.: ENVIRON Corporation.

Roe, E. M. 1988. A case study of the 1980/82 Medfly controversy in California. Working paper prepared for the National Research Council's Committee on Risk Perception and Communication.

Rosenblatt, R. A. 1976. GAO Calls U.S. A-Power Booklet 'Propaganda.' Los Angeles Times, September 30, 1976, I-30.

Ross, L., and C. A. Anderson. 1982. Shortcomings in the attribution process: On the origins and maintenance of erroneous social assessments. In Judgment Under Uncertainty: Heuristics and Biases, D. Kahneman, P. Slovic, and A. Tversky, eds. New York: Cambridge University Press.

Ruckelshaus, W. D. 1983. Science, risk and public policy. Science 221:1026–1028.

Ruckelshaus, W. D. 1984. Risk in a Free Society. Risk Analysis 4(3):157–162.

Ruckelshaus, W. D. 1985. Risk, science, and democracy. Issues in Science and Technology 1(3):19–38.

Ruckelshaus, W. D. 1987. Communicating about risk. In Risk Communication, J. C. Davies, V. T. Covello, and F. W. Allen, eds. Washington, D.C.: The Conservation Foundation.

Rushefsky, M. E. 1984. The misuse of science in governmental decisionmaking. Science, Technology, and Human Values 9(3):47–59.

Russell, M., and M. Gruber. 1987. Risk assessment in environmental policymaking. Science 236(4799):286-290.

Sagan, L. A. 1987. Beyond risk assessment. Risk Analysis 7(1):1–2.

Sandman, P. M. 1986. Explaining Environmental Risk: Some Notes on Environmental Risk Communication. Washington, D.C.: U.S. Environmental Protection Agency.

Sandman, P. M., D. B. Sachsman, and M. R. Greenberg. 1987a. The Environmental News Source: Informing the Media During an Environmental Crisis. New Brunswick, N.J.: Rutgers University.

Sandman, P. M., N. D. Weinstein, and M. L. Klotz. 1987b. Public response to the risk from geological radon. Journal of Communication 37(3):93–108.

Sapolsky, H. M. 1968. Science, voters, and the fluoridation controversy. Science 162(October 25):427–433.

Sapolsky, H. M., ed. 1986. Consuming Fears: The Politics of Product Risks. New York: Basic Books.

Sapolsky, H. M. 1987. Is honesty the best policy? In AIDS Public Policy Dimensions, J. Griggs, ed. New York: United Hospital Fund of New York.

Scenic Hudson Preservation Conference v. *Federal Power Commission,* 354 F. 2d 608 (2d Cir. 1965).

Schmandt, J. 1984. Regulation and science. Science, Technology, and Human Values 9(1):23–38.

Schneiderman, M. A. 1983. Cancer: Scientific policy, public policy, and the prevention of disease. Carolina Environmental Essay Series. Chapel Hill, N.C.: Institute for Environmental Studies.

Schneiderman, M. A. 1987a. Risk assessment—where do we want it to go? In Quantitative Risk Assessment: Biomedical Ethics Reviews, J. M. Humber and R. F. Almeder, eds. Clifton, N.J.: Humana Press.

Schneiderman, M. A. 1987b. Expectation and limitation of human studies and risk assessment. In Health Effects from Hazardous Waste Sites, J. B. Andelman and D. W. Underhill, eds. Chelsea, Mich.: Lewis Publishers.

Schudson, M. 1978. Discovering the News. New York: Basic Books.

Schultz, W., G. McClelland, B. Hurd, and J. Smith. 1986. Improving Accuracy and Reducing Costs of Environmental Benefits Assessment. Vol. IV. Boulder: University of Colorado, Center for Economic Analysis.

Schuman, H., and J. Scott. 1987. Problems in the use of survey questions to measure public opinion. Science 236(4804):957–959.

Schwartz, S. H. 1977. Normative influences on altruism. In Advances in Experimental Social Psychology, Vol. 10, L. Berkowitz, ed. New York: Academic Press.

Sharlin, H. I. 1987. Macro-risks, micro-risks, and the media: The EDB case. In The Social and Cultural Construction of Risk, B. B. Johnson and V. T. Covello, eds. Dordrecht, Holland: D. Reidel.

Sherry, S. 1985. High Tech and Toxics: A Guide for Local Communities. Washington, D.C.: National Center for Policy Alternatives.

Sheth, J. N., and G. L. Frazier. 1982. A model of strategy mix choice or planned social change. Journal of Marketing 46:15–26.

Shilts, R. 1987. And the Band Played On: Politics, People, and the AIDS Epidemic. New York: St. Martin's Press.

Short, J. F., Jr. 1984. The social fabric at risk: Toward the social transformation of risk analysis. American Sociological Review 49:711.

Siegel, B. 1987. "Managing" Risks: Sense and Science. Los Angeles Times, July 5:Il.

Silbergeld, E. 1987a. Responsibilities of risk communicators. In Risk Communication, J. C. Davies, V. T. Covello, and F. W. Allen, eds. Washington, D.C.: The Conservation Foundation.

Silbergeld, E. 1987b. Letter. Science 237(4821):1399.

Simmons, J. 1984. Rights and wrongs in hazardous waste disposal. In Not in My Backyard! Community Reaction to Locally Unwanted Land Use. Charlottesville: University of Virginia.

Sindell v. Abbott Laboratories, 26 Cal. 3d 588 (1980).

Singer, S., and P. Endreny. 1987. Reporting hazards: Their benefits and costs. Journal of Communication 37(3):10–26.

Slovic, P. 1986. Informing and educating the public about risk. Risk Analysis 6(4):403–415.

Slovic, P. 1987. Perception of risk. Science 236(4799):280–285.

Slovic, P., B. Fischhoff, and S. Lichtenstein. 1979. Rating the risks. Environment 21(3):14–20, 36–39.

Slovic, P., B. Fischhoff, and S. Lichtenstein. 1980. Perceived risk. In Societal Risk Assessment: How Safe Is Safe Enough?, R. Schwing and W. A. Albers, eds. New York: Plenum Press.

Slovic, P., S. Lichtenstein, and B. Fischhoff. 1988. Decision making. In Stevens' Handbook of Experimental Psychology. New York: John Wiley & Sons.

Smith, H. L. 1974. Myocardial infarction—case studies of ethics in the consent situation. Social Science and Medicine 8:399–404.

Smith, L. 1987. STOP I.T.: Citizens' response to hazardous waste treatment facility. Presented to National Research Council Committee on Risk Perception and Communication meeting, Washington, D.C., July 23, 1987.

Smith, R. J. 1983. Covering the EPA: Or, wake me up if anything happens. Columbia Journalism Review 22(September–October):29–34.

Smith, R. J. 1986. Chernobyl report surprisingly detailed but avoids painful truths, experts say. Washington Post, August 27:A25.

Smith, V. K., W. H. Desvousges, A. Fisher, and F. R. Johnson. 1987. Communicating radon risk effectively: A mid-course evaluation. EPA Cooperative Agreement No. CR-811075. Washington, D.C.: U.S. Environmental Protection Agency, Office of Policy Analysis.

Sood, R., G. Stockdale, and E. M. Rogers. 1987. How the news media operate in natural disasters. Journal of Communications 37(3):27–41.

Sosenko, A. 1983. After Burford, what? The time has come for a new EPA beginning. Inside EPA 4(11):3.

Spain, D. 1984. Women's role in opposing locally unwanted land uses. In Not in My Backyard! Community Reaction to Locally Unwanted Land Use. Charlottesville: University of Virginia.

Stallen, P. J., and R. Coppock. 1987. About risk communication and risky communication. Risk Analysis 7(4):413–414.

Starr, C., and C. Whipple. 1980. Risks of risk decisions. Science 208(4448):1114–1119.

Starr, P. 1982. The Social Transformation of American Medicine. New York: Basic Books.

Stephens, M., and N. G. Edison. 1982. News media coverage of issues during the accident at Three Mile Island. Journalism Quarterly 59:199–204.

Stern, P. C., and E. Aronson. 1984. Energy use: The human dimension. New York: W. H. Freeman.

Stern, P. C., T. Dietz, and J. S. Black. 1986. Support for environmental protection: The role of moral norms. Population and Environment 8:204–222.

Suplee, C. 1987. Semiotics: In search of more perfect persuasion. The Washington Post, January 18:C3.

Svenson, O., and B. Fischhoff. 1985. Levels of environmental decisions. Journal of Environmental Psychology 5:55–67.

Tarr, J. A., and C. Jacobson. 1987. Environmental risk in historical perspective. In The Social and Cultural Construction of Risk, B. B. Johnson and V. T. Covello, eds. Dordrecht, Holland: D. Reidel.

Thaler, R. 1985. Mental accounting and consumer choice. Marketing Science 4(3):199-214.

Thomas, L. 1987. Making and communicating pesticide decisions. EPA Journal 13(4):3.

Tichenor, P., G. Donohue, and C. Olien. 1980. Community Conflict and the Press. Beverly Hills, Calif.: Sage.

Tiemann, A. R. 1987. Comment: Risk, technology, and society. Risk Analysis 7(1):11–13.

Tierney, J. 1988. Not to worry. Hippocrates (January/February):29–38.

Touraine, A., et al. 1983. Anti-nuclear Protest: The Opposition to Nuclear Energy in France. New York: Cambridge University Press.

Towle, M. 1986. Carbide neighbors view leak as catalyst for change. Charleston Daily Mail, August 8:1-A.

Travis, C. C., and R. K. White. 1988. Interspecific scaling of toxicity data. Risk Analysis 8(1):119–125.

Tversky, A., and D. Kahneman. 1971. The belief in the "law of small numbers." Psychological Bulletin 76:105–110.

Tversky, A., and D. Kahneman. 1973. Availability: A heuristic for judging frequency and probability. Cognitive Psychology 5:207–232.

Tversky, A., and D. Kahneman. 1981. The framing of decisions and the rationality of choice. Science 211(4481):453–458.

Urquhart, J., and K. Heilmann. 1984. Riskwatch: The Odds of Life. New York: Facts on File.

U.S. Bureau of the Census. 1980. Social Indicators III. Washington, D.C.: U.S. Bureau of the Census.

U.S. Committee on Government Operations. 1978. Teton Dam Disaster. Washington, D.C.: U.S. Government Printing Office.

U.S. Congress, Senate. 1987. S.638, A bill to require the Secretary of Health and Human Services to develop standards governing the notification of individuals who have been exposed to hazardous substances or physical agents, and for other purposes. 100th Cong., 1st sess.

Verba, S., and N. H. Nie. 1972. Participation in America: Political Democracy and Social Equality. New York: Harper & Row.

Viscusi, W. K. 1985. A Bayesian perspective on biases on risk perception. Economics Letters 17:59–62.

Viscusi, W. K., and C. J. O'Connor. 1984. Adaptive responses to chemical labeling: Are workers Bayesian decision makers? The American Economic Review 74(5):942–956.

Viscusi, W. K., W. A. Magat, and J. Huber. 1986. Informational regulation of consumer health risks: An empirical evaluation of hazard warnings. Rand Journal of Economics 17(3):351–365.

Viscusi, W. K., W. A. Magat, and J. Huber. 1987. An investigation of the rationality of consumer valuations of multiple health risks. Rand Journal of Economics 18(4):465–479.

von Winterfeldt, D., and W. Edwards. 1984. Patterns of conflict about risky technologies. Risk Analysis 4(1):55–68.

von Winterfeldt, D., and W. Edwards. 1986. Decision Analysis and Behavioral Research. New York: Cambridge University Press.

von Winterfeldt, D., R. S. John, and K. Borcherding. 1981. Cognitive components of risk ratings. Risk Analysis 1(4):277–287.

Wallsten, T., and D. Budescu. 1983. Encoding subjective probabilities: A psychological and psychometric review. Management Science 29:135–140.

Weinstein, N. D., P. M. Sandman, and M. L. Klotz. 1987. Public response to the risk from radon, 1986. Research Contract C29543. Trenton: Division of Environmental Quality, New Jersey Department of Environmental Protection.

Whittemore, A. S. 1983. Facts and values in risk analysis for environmental toxicants. Risk Analysis 3(1):23–33.

Wildavsky, A. 1988. Searching for Safety. New Brunswick, N.J.: Transaction Books.

Wilkie, W. L., and D. M. Gardner. 1974. The role of marketing research in public policy decision making. Journal of Marketing 38:38–47.

Wilkins, L., and P. Patterson. 1987. Risk analysis and the construction of news. Journal of Communication 37(3):80–92.

Wilkinson, C. F. Undated. Communicating with the Public and the Media. Washington, D.C.: Institute for Health Policy Analysis, Georgetown University Medical Center.

Wilson, R., and E. A. C. Crouch. 1987a. Risk assessment and comparisons: An introduction. Science 236(4799):267–270.

Wilson, R., and E. A. C. Crouch. 1987b. Letter. Science 237(4821):1400.

Wilson, R., S. D. Colome, J. D. Spengler, and D. G. Wilson. 1980. Health Effects of Fossil Fuel Burning: Assessment and Mitigation. Cambridge, Mass.: Ballinger.

Wise, J. 1987. Risk management in a climate of public fear. Paper presented at Peninsula Industrial and Business Association meeting, June 12, 1987.

Wolfe, A. K. 1986. Confidence in Technologies: Interactions Between Publics and Industries. Paper presented at the Society for Risk Analysis Annual Meeting, Boston, Mass., November 11.

Zeckhauser, R., and D. S. Shephard. 1981. Principles for saving and valuing lives. In The Benefits of Health and Safety Regulation, A. Ferguson and E. P. LeVeen, eds. Cambridge, Mass.: Ballinger.

Zimmerman, R. 1987. A process framework for risk communication. Science, Technology, and Human Values 12(3&4):131–137.

Appendix C
Risk: A Guide to Controversy

BARUCH FISCHHOFF

FOREWORD BY THE COMMITTEE

This appendix was written by Baruch Fischhoff to assist in the deliberations of the National Research Council's Committee on Risk Perception and Communication. It describes in some detail the complications involved in controversies over managing risks in which risk perception and risk communication play significant roles. It addresses these issues from the perspective of many years of research in psychology and other disciplines. The text of the committee's report addresses many of the same issues, and, not surprisingly, many of the same themes, although the focus of the report is more general. The committee did not debate all points made in the guide. Even though this appendix represents the views of only one member, the committee decided to include it because we believe the guide to be a valuable introduction to an extremely complicated literature.

PREFACE

This guide is intended to be used as a practical aid in applying general principles to understanding specific risk management controversies and their associated communications. It might be thought of as a user's guide to risk. Its form is that of a "diagnostic guide," showing participants and observers how to characterize risk controversies

211

along five essential dimensions, such as "What are the (psychological) obstacles to laypeople's understanding of risks?" and "What are the limits to scientific estimates of riskiness?" Its style is intended to be nontechnical, thereby making the scientific literature on risk accessible to a general audience. It is hoped that the guide will help make risk controversies more comprehensible and help citizens and professional risk managers play more effective roles in them.

The guide was written for the committee by one of its members. Its substantive contents were considered by the committee in the course of its work, either in the form of published articles and books circulated to other committee members or in the form of issues deliberated at its meetings. As a document, the guide complements the conclusions of the committee's report.

CONTENTS

I
INTRODUCTION

Risk management is a complex business. So are the controversies that it spawns. And so are the roles that risk communication must perform. In the face of such complexity, it is tempting to look for simplifying assumptions. Made explicit, these assumptions might be expressed as broad statements of the form, "what people really want is . . ."; "all that laypeople can understand is . . ."; or "industry's communicators fail whenever they. . . ." Like other simplifications in life, such assumptions provide some short-term relief at the price of creating long-term complications. Overlooking complexities eventually leads to inexplicable events and ineffective actions.

On one level this guide might be used like a baseball scorecard detailing the players' identities and performance statistics (perhaps along with any unique features of the stadium, season, and rivalry). Like a ballgame, a risk controversy should be less confusing to spectators who know something about the players and their likely behavior under various circumstances. Thus, experts might respect the public more if they were better able to predict its behavior, even if they would prefer that the public behave otherwise. Similarly, understanding the basics of risk analysis might make disputes among technical experts seem less capricious to the lay public.

More ambitiously, such a guide might be used to facilitate effective action by the parties in risk controversies, like the *Baseball Abstract* (James, 1988) in the hands of a skilled manager. For example, the guide discusses how to determine what the public needs to know in particular risky situations. Being able to identify those needs may allow better focused risk communication, thereby using the public's limited time wisely and letting it know that the communicators really care about the problems that the public faces. Similarly, understanding the ethical values embedded in the definitions of ostensibly technical terms (e.g., risk, benefit, voluntary) can allow members of the public to ask more penetrating questions about whose interests a risk analysis serves. Realizing that different actors use a term like "risk" differently should allow communicators to remove that barrier to mutual understanding.

214

USAGE

The guide's audience includes all participants and observers of risk management episodes involving communications. Its intent is to help government officials preparing to address citizens' groups, industry representatives hoping to site a hazardous facility without undue controversy, local activists trying to decide what information they need and whether existing communications meet those needs, and academics wondering how central their expertise is to a particular episode.

The premise of the guide is that risk communication cannot be understood in isolation. Rather, it is one component of complex social processes involving complex individuals. As a result, this fuller context needs to be understood before risk communication can be effectively transmitted or received. That context includes the following elements and questions:

• *The Science.* What is the scientific basis of the controversy? What kinds of risks and benefits are at stake? How well are they understood? How controversial is the underlying science? Where does judgment enter the risk estimation process? How well is it to be trusted?

• *Science and Policy.* In what ways does the nature of the science preempt the policymaking process (e.g., in the definition of key terms, like "risk" and "benefit"; in the norms of designing and reporting studies)? To what extent can issues of fact and of value be separated?

• *The Nature of the Controversy.* Why is there a perceived need for risk communication? Does the controversy reflect just a disagreement about the magnitude of risks? Is controversy over risk a surrogate for controversy over other issues?

• *Strategies for Risk Communication.* What are the goals of risk communication? How can communications be evaluated? What burden of responsibility do communicators bear for evaluating their communications, both before and after dissemination? What are the alternatives for designing risk communication programs? What are the strengths and weaknesses of different approaches? How can complementary approaches be combined? What nonscientific information is essential (e.g., the mandates of regulatory agencies, the reward schemes of scientists)?

• *Psychological Principles in Communication Design.* What are the behavioral obstacles to effective risk communication? What kinds of scientific results do laypeople have difficulty understanding? How does emotion affect their interpretation of reported results? What presentations exacerbate (and ameliorate) these problems? How does personal experience with risks affect people's understanding?

SOME CAUTIONS

A diagnostic guide attempts to help users characterize a situation. To do so, it must define a range of possible situations, only one of which can be experienced at a particular time. As a result, the attempt to make one guide fit a large universe of risk management situations means that readers will initially have to read about many potential situations in order to locate the real situation that interests them. With practice, users should gain fluency with a diagnostic approach, making it easier to characterize specific situations. It is hoped that the full guide will be interesting enough to make the full picture seem worth knowing.

At no time, however, will diagnosis be simple or human behavior be completely predictable. All that this, or any other, diagnostic guide can hope to do is ensure that significant elements of a social-political-psychological process are not overlooked. For a more detailed treatment, one must look to the underlying research literature for methods and results. To that end, the guide provides numerous references to that literature, as well as some discussion of its strengths and limitations.

To the extent that a guide is useful for designing and interpreting a communication process, it may also be useful for manipulating that process. In this regard, the material it presents is no different than any other scientific knowledge. This possibility imposes a responsibility to make research equally available to all parties. Therefore, even though this guide may suggest ways to bias the process, it should also make it easier to detect and defuse such attempts.

II
THE SCIENCE

By definition, all risk controversies concern the risks associated with some hazard. However, as argued in the text of the report and in this diagnostic guide, few controversies are only about the size of those risks. Indeed, in many cases, the risks prove to be a side issue, upon which are hung disagreements about the size and distribution of benefits or about the allocation of political power in a society. In all cases, though, some understanding of the science of risk is needed, if only to establish that a rough understanding of the magnitude of the risk is all that one needs for effective participation in the risk debate. Following the text, the term "hazard" is used to describe any activity or technology that produces a risk. This usage should not obscure the fact that hazards often produce benefits as well as risks.

Understanding the science associated with a hazard requires a series of essential steps. The first is identifying the scope of the problem under consideration, in the sense of identifying the set of factors that determine the magnitude of the risks and benefits produced by an activity or technology. The second step is identifying the set of widely accepted scientific "facts" that can be applied to the problem; even when laypeople cannot understand the science underlying these facts, they may at least be able to ensure that such accepted wisdom is not contradicted or ignored in the debate over a risk. The third step in understanding the science of risk is knowing how it depends on the educated intuitions of scientists, rather than on accepted hard facts; although these may be the judgments of trained experts, they still need to be recognized as matters of conjecture that are both more likely to be overturned than published (and replicated) results and more vulnerable to the vagaries of psychological processes.

WHAT ARE THE BOUNDS OF THE PROBLEM?

The science learned in school offers relatively tidy problems. The typical exercise in, say, physics gives all the facts needed for its solution and nothing but those facts. The difficulty of such problems for students comes in assembling those facts in a way that provides the right answer. (In more advanced classes, one may have to bring some general facts to bear as well.)

The same assembly problem arises when analyzing the risks and benefits of a hazard. Scientists must discover how its pieces fit together. They must also figure out what the pieces are. For example, what factors can influence the reliability of a nuclear power plant? Or, whose interests must be considered when assessing the benefits of its operation? Or, which alternative ways of generating electricity are realistic possibilities?

The scientists responsible for any piece of a risk problem must face a set of such issues before beginning their work. Laypeople trying to follow a risk debate must understand how various groups of scientists have defined their pieces of the problem. And, as mentioned in the report, even the most accomplished of scientists are laypeople when it comes to any aspects of a risk debate outside the range of their trained expertise.

The difficulties of determining the scope of a risk debate emerge quite clearly when one considers the situation of a reporter assigned to cover a risk story. The difficult part of getting most environmental stories is that no one person has the entire story to give. Such stories typically involve diverse kinds of expertise so that a thorough journalist might have to interview specialists in toxicology, epidemiology, economics, groundwater movement, meteorology, and emergency evacuation, not to mention a variety of local, state, and federal officials concerned with public health, civil defense, education, and transportation.

Even if a reporter consults with all the relevant experts, there is no assurance of complete coverage. For some aspects of some hazards, no one may be responsible.

For example, no evacuation plans may exist for residential areas that are packed "hopelessly" close to an industrial facility. No one may be capable of resolving the jurisdictional conflicts when a train with military cargo derails near a reservoir just outside a major population center. There may be no scientific expertise anywhere for measuring the long-term neurological risks of a new chemical.

Even when there is a central address for questions, those occupying it may not be empowered to take firm action (e.g., banning or exonerating a chemical) or to provide clear-cut answers to personal questions (e.g., "What should I do?" or "What should I tell my children?"). Often those who have the relevant information refuse to divulge it because it might reveal proprietary secrets or turn public opinion against their cause.

Having to piece together a story from multiple sources, even recalcitrant ones, is hardly new to journalists. What is new about many environmental stories is that no one knows what all of the pieces are or realizes the limits of their own understanding.

Experts tend to exaggerate the centrality of their roles. Toxicologists may assume that everyone needs to know what they found when feeding rats a potential carcinogen or when testing groundwater near a landfill, even though additional information is always needed to make use of those results (e.g., physiological differences among species, routes of human exposure, compensating benefits of the exposure).

Another source of confusion is the failure of experts to remind laypeople of the acknowledged limits of the experts' craft. For example, cost-benefit analysts seldom remind readers that the calculations consider only total costs and benefits and, hence, ignore questions of who pays the costs and who pays the benefits (Bentkover et al., 1985; Smith and Desvousges, 1986).

Finally, environmental management is an evolving field that is only beginning to establish comprehensive training programs and methods, making it hard for anyone to know what the full picture is and how their work fits into it.

An enterprising journalist with a modicum of technical knowledge should be able to get specialists to tell their stories in fairly plain English and to cope with moderate evasiveness or manipulation. However, what is the journalist to do when the experts do not know what they do not know? One obvious solution is to talk to several experts with maximally diverse backgrounds. Yet, sometimes such a perfect mix is hard to find. Available experts can all have common limitations of perspective.

Another solution is to use a checklist of issues that need to be covered in any comprehensive environmental story. Scientists themselves use such lists to ensure that their own work is properly performed, documented, and reported. Such a protocol does not create knowledge for the expert any more than it would provide an education to the journalist. It does, however, help users exploit all they know—and acknowledge what they leave out.

Some protocols that can be used in looking at risk analyses are the causal model, the fault tree, a materials and energy flow diagram, and a risk analysis checklist.

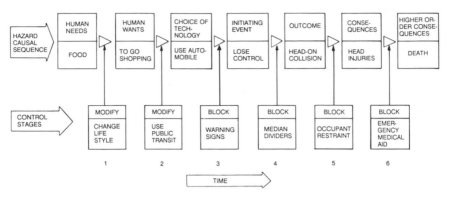

FIGURE II.1   The causal chain of hazard evolution. The top line indicates seven stages of hazard development, from the earliest (left) to the final stage (right). These stages are expressed generically in the top of each box and in terms of a sample motor vehicle accident in the bottom. The stages are linked by causal pathways denoted by triangles. Six control stages are linked to pathways between hazard states by vertical arrows. Each is described generically as well as by specific control actions. Thus control stage 2 would read: "You can modify technology choice by substituting public transit for automobile use and thus block the further evolution of the motor vehicle accident sequence arising out of automobile use." The time dimension refers to the ordering of a specific hazard sequence; it does not necessarily indicate the time scale of managerial action. Thus, from a managerial point of view, the occurrence of certain hazard consequences may trigger control actions that affect events earlier in the hazard sequence. SOURCE: Figure—Bick et al., 1979; caption—Fischhoff, Lichtenstein, et al., 1981.

The Causal Model

The causal model of hazard creation is a way to organize the full set of factors leading to and from an environmental mishap, both when getting the story and when telling it. The example in Figure II.1 is an automobile accident, traced from the need for transportation to the secondary consequence of the collision. Between each stage, there is some opportunity for an intervention to reduce the risk of an accident. By organizing information about the hazard in a chronological sequence, this scheme helps ensure that nothing is left out, such as the deep-seated causes of the mishap (to the left) and its long-range consequences (to the right).

Applied to an "irregular event" at a nuclear power station, for example, this protocol would work to remind a reporter of such (left-handed) causes as the need for energy and the need to protect the large capital investment in that industry and such (right-handed) consequences as the costs of retooling other plants designed like the affected plant or the need to burn more fossil fuels if the plant is taken off line (without compensating reductions in energy consumption).

The Fault Tree

A variant on this procedure is the fault tree (Figure II.2), which lays out the sequence of events that must occur for a particular accident to happen (Green and Bourne, 1972; U.S. Nuclear Regulatory Commission, 1983). Actual fault trees, which can be vastly more involved than this example, are commonly used to organize the thinking and to coordinate the work of those designing complex technologies such as nuclear power facilities and chemical plants. At times, they are also used to estimate the overall riskiness of such facilities. However, the numbers produced are typically quite imprecise (U.S. Nuclear Regulatory Commission, 1978).

In effect, fault trees break open the right-handed parts of a causal model for detailed treatment. They can help a reporter to

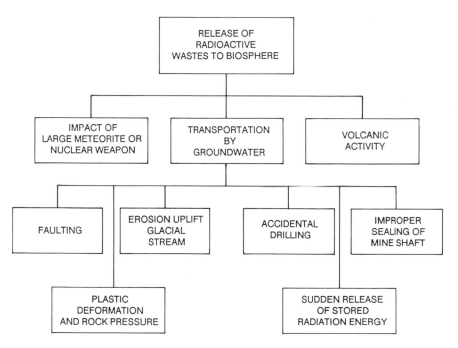

FIGURE II.2 Fault tree indicating the possible ways that radioactivity could be released from deposited wastes after the closure of a repository. SOURCE: Slovic and Fischhoff, 1983.

order the pieces of an accident story collected from different sources, see where an evolving incident (e.g., Three Mile Island or a leaking waste dump) is heading, and find out what safety measures were or were not taken.

Materials and Energy Flow Diagrams

The next model (Figure II.3) is adapted from the engineering notion of a materials or energy flow diagram. If something is neither created nor destroyed in a process, then one should be able to account schematically for every bit of it. In environmental affairs, one wants to account for all toxic materials. It is important to know where each toxic agent comes from and where each goes.

Keeping track of a substance can help anticipate where problems will appear, recur, and disappear. It can reveal when a problem has actually been treated and when it has merely been shifted to another time, place, or jurisdiction. With a story like EDB (ethylene dibromide, a fungicide used on grain) (Sharlin, 1987), such a chart would have encouraged questions such as, does it decay with storage or does it become something even worse when cooked and digested? Applying this approach led Harriss and Hohenemser (1978) to conclude that pollution controls had not reduced the total amount of mercury released into the environment, but only the distribution of releases (replacing a few big polluters with many smaller ones). In creating such figures, it is important to distinguish between where a substance is supposed to go and where it actually goes.

A comparable figure might be drawn to keep track of where the money goes, identifying the beneficiaries and losers resulting from different regulatory actions. With the EDB story, such a chart would have encouraged questions about who would eventually pay for the grain lost to pests if that chemical were not used. That is, would reducing the risk of EDB reduce producers' profits or increase consumers' prices? In the former case, failure to ban EDB looks much more callous than in the latter.

A Risk Analysis Checklist

The fourth aid (Figure II.4) is a list of questions that can be asked in a risk analysis (or of a risk analyst) in order to clarify what problem has been addressed and how well it has been solved.

This list was compiled for a citizens' group concerned with pesticides. Its members had mastered many substantive details of the

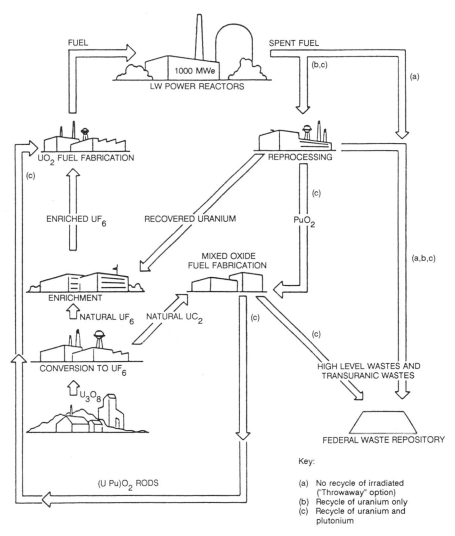

FIGURE II.3 Materials and energy flow diagram: Current options for the nuclear fuel cycle. SOURCE: Gotchy, 1983.

discipline, such as toxicology and biochemistry, involved in pesticide management, when suddenly they were confronted with a new procedure—risk analysis. In principle, risk analysis does no more than organize information from substantive disciplines in a way that allows overall estimates of risk to be computed. It can facilitate citizen access by forcing all the facts out on the table.

1. Does the risk analysis <u>first</u> state the health damage that may occur and <u>then</u> present the odds (i.e., the risk analysis)?
2. Is there enough information available on the factors that are most crucial to risk calculations?
3. If some of the data are missing, but there are enough to <u>approach</u> a risk assessment, are the missing data labeled as such?
4. Does the risk analysis disclose forthrightly the points at which it is based on guesswork?
5. Are various risk factors allowed to assume a variety of numbers depending on uncertainties in the data and/or various interpretations of the data?
6. Does the risk analysis multiply its probabilities by the number of people exposed to produce the number of people predicted to suffer damage?
7. Does the risk analysis disclose the confidence limits for its projections and the method of arriving at those confidence limits?
8. Are considerations of individual sensitivities, exposure to multiple pesticides, cumulative effects, and effects other than cancer, birth defects, and mutations included in the risk analysis?
9. Are all data and processes of the risk analysis open to public scrutiny?
10. Has an independent peer review of the risk analysis been funded and made public?
11. Are questions of (a) involuntary exposure, (b) who bears the risks and who reaps the benefits, and (c) alternatives to pesticide use being considered alongside the risk analysis?
12. Are alternatives to pesticide use also being extensively analyzed for risk or lack of risk?
13. Are the processes of risk analysis and risk policy separate?

FIGURE II.4   Risk analysis checklist. SOURCE: Northwest Coalition for Alternatives to Pesticides, 1985.

However, unless one can penetrate all its formalisms, risk analysis can mystify and obscure the facts rather than reveal them. Such a checklist can clarify what an analysis has done in terms approximating plain English.

WHAT IS THE HARD SCIENCE RELATED TO THE PROBLEM?

With most "interesting" hazards, the data run out long before enough is known to estimate their risks and benefits as precisely as one would want. Much of risk management involves going beyond the available data either to guess at what the facts might be or to figure out how to live with uncertainty. Obviously, one wants to reduce this uncertainty by making the best of the hard data available.

Unfortunately, there is no short-cut to providing observers with ways to read critically all of the kinds of science that could be invoked in the course of characterizing a risk. There are too many sciences to consider and too many nuances in each type of science to know about in assessing the validity of studies conducted in any one field. Even the social sciences, which seem relatively accessible (compared with the physical sciences) and the results of which can be rendered into common English, routinely foil the efforts of amateur scientists.

These failures can be seen most clearly in the attempts by non-social scientists to make factual statements about the behavior of laypeople, solely on the basis of their untrained anecdotal observations. Such speculations can mislead more than inform if they are made without realizing that they lack the discipline of science.

The complexities of science arise in the details of creating, analyzing, and interpreting specific sets of data. To give a feeling for these strengths and limits of scientific research, several examples drawn from social science research into risk perception and communication are presented here. Each science has its own nuances. Featuring this science also provides background for interpreting the social science results described below.

Like speculations about chemical reactions, speculations about human behavior must be disciplined by fact. Such speculations make important statements about people and their capabilities, and failure to validate them may mean arrogating to oneself considerable political power. Such happens, for example, when one says that people are so poorly informed (and ineducable) they require paternalistic institutions to defend them, and, furthermore, they might be better off surrendering some political rights to technical experts. It also happens, at the other extreme, when one claims that people are so well informed (and offered such freedom of choice) one need not ask them anything at all about their desires; to know what they want, one need only observe their behavior in the marketplace. It also happens when we assume that people are consummate hedonists, rational to the extreme in their consumer behavior but totally uncomprehending of broader economic issues, so we can impose effective fiscal policies on them without being second-guessed.

One reason for the survival of such simplistic and contradictory positions is political convenience. Some people want the lay public to participate actively in hazard management decisions, and need to be able to describe the public as competent; others need an incompetent public to legitimate an expert elite. A second reason is theoretical convenience. It is hard to build models of people who are sometimes wise and sometimes foolish, sometimes risk seeking and sometimes risk averse. A third reason is that one can effortlessly speculate about human nature and even produce a bit of supporting anecdotal information. Indeed, good social theory may be so rare because poor social theory is so easy.

Judgments of Risk

At first sight, assessing the public's risk perceptions would seem to be very straightforward. Just ask questions like, "What is the probability of a nuclear core meltdown?" or "How many people die annually from asbestos-related diseases?" or "How does wearing a seat belt affect your probability of living through the year?" Once the results are in, they can be compared with the best available technical estimates, with deviations interpreted as evidence of respondents' ignorance.

Unfortunately, how one asks the question may in large part determine the content (and apparent wisdom) of the response. Lichtenstein and her colleagues (Lichtenstein et al., 1978) asked two groups of educated laypeople to estimate the frequency of death in the United States from each of 40 different causes. The groups differed only in the information that was given to them about one cause of death in order to help scale their responses. One group was told about 50,000 people die annually in motor vehicle accidents, and the other was told about 1,000 annual deaths result from electrocution. Both reports were accurate, but receiving a larger number increased the estimates of most frequencies for respondents in the motor vehicle accident group. This is a special case of a general psychological phenomenon called "anchoring," whereby people's responses are pulled toward readily available numbers in cases in which they do not know exactly what to say (Poulton, 1968, 1977; Tversky and Kahneman, 1974). Such anchoring on the original number changed the smallest estimates by roughly a factor of 5.

Fischhoff and MacGregor (1983) asked people to judge the lethality of various potential causes of death using one of four formally equivalent formats (e.g., "For each afflicted person who dies, how many survive?" or "For each 100,000 people afflicted, how many will die?"). Table II.1 expresses their judgments in a common format and reveals even more dramatic effects of question phrasing on expressed risk perceptions. For example, when people estimated the lethality rate for influenza directly (column 1), their mean response was 393 deaths per 100,000 cases. When told that 80 million people catch influenza in a normal year and asked to estimate the

TABLE II.1 Lethality Judgments with Four Different Response Modes
(geometric mean)

Condition	Death Rate Per 100,000 Afflicted				
	Estimated Lethality Rate	Estimated Number Who Die	Estimated Survival Rate	Estimated Number Who Survive	Actual Lethality Rate
Influenza	393	6	26	511	1
Mumps	44	114	19	4	12
Asthma	155	12	14	599	33
Venereal disease	91	63	8	111	50
High blood pressure	535	89	17	538	76
Bronchitis	162	19	43	2111	85
Pregnancy	67	24	13	787	250
Diabetes	487	101	52	5666	800
Tuberculosis	852	1783	188	8520	1535
Automobile accidents	6195	3272	31	6813	2500
Strokes	11,011	4648	181	24,758	11,765
Heart attacks	13,011	3666	131	27,477	16,250
Cancer	10,889	10,475	160	21,749	37,500

NOTE: The four experimental groups were given the following instructions:
(a) Estimate lethality rate: For each 100,000 people afflicted, how many die?
(b) Estimate number who die: X people were afflicted, how many died?
(c) Estimate survival rate: For each person who died, how many were afflicted but survived?
(d) Estimate number who survive: Y people died, how many were afflicted but did not die?
Responses to (b), (c), and (d) were converted to deaths per 100,000 to facilitate comparisons.

SOURCE: Fischhoff and MacGregor, 1983.

number who die (column 2), their mean response was 4800, representing a death rate of only 6 per 100,000 cases. This slight change in the question changed the estimated rate by a factor of more than 60. Similar discrepancies occurred with other questions and other hazards. One consequence for risk communicators is that whether laypeople intuitively overestimate or underestimate risks (or perceive them accurately) depends on what question they are asked.

In a recent study at an Ivy League college (Linville et al., 1988), students were asked to give estimates of the probability that the AIDS virus could be transmitted from a man to a woman in a single case of unprotected sex. The median estimate was about 10 percent, considerably above current scientific estimates (Fineberg,

1988). However, when asked to give estimates for the probability of transmission in 100 cases of unprotected sex, the median answer was about 25 percent. This risk estimate is considerably more in line with scientific thinking—so that an investigator asking this question would have a considerably more optimistic assessment of the state of public understanding. Unfortunately, it is also completely inconsistent with the single-case estimates produced by the same individuals. If one believes in a single-case probability of 10 percent, then transmission should be a virtual certainty with 100 exposures. Such failure to see how small risks mount up over repeated exposures has been observed in such diverse settings as the risks from playing simple gambles (Bar-Hillel, 1973), driving (Slovic et al., 1978), and relying on various contraceptive devices (Shaklee et al., 1988).

Such effects are hardly new; indeed, some have been recognized for close to 100 years. Early psychologists discovered that different numerical judgments may be attached to the same physical stimulus (e.g., the loudness of a tone) as a function of whether the set of alternatives is homogeneous or diverse, and whether the respondent makes one or many judgments. Even when the same presentation is used, different judgments might be obtained with a numerical or a comparative (ordinal) response mode, with instructions stressing speed or accuracy, with a bounded or an unbounded response set, and with verbal or numerical response labels.

The range of these effects may suggest that the study of judgment is not just difficult, but actually impossible. Closer inspection, however, reveals considerable orderliness underlying this apparent chaos (Atkinson et al., 1988; Carterette and Friedman, 1974; Woodworth and Schlosberg, 1954).

Judgments of Values

Once the facts of an issue have been estimated and communicated, it is usually held that laypeople should (in a democracy) be asked about their values. What do they want—after the experts have told them what they can (conceivably) have? Here, too, the straightforward strategy of "just ask them" runs into trouble.

The problem of poorly (or even misleadingly) worded questions in attitude surveys is well known, although not necessarily well resolved (Bradburn and Sudman, 1979; National Research Council, 1982; Payne, 1952; Zeisel, 1980). For example, a major trade publication (Ventner, 1979) presented the results of a survey of public attitudes toward the chemical industry containing the following question:

> Some people say that the prime responsibility for reducing exposure of workers to dangerous substances rests with the workers themselves, and that all substances in the workplace should be clearly labeled as to their levels of danger and workers then encouraged or forced to be careful with these substances. Do you agree or disagree?

It is hard to know what one is endorsing when one says "Yes," "No," or "I don't know" to such a complex and unclear question.

Although annoying, ambiguous wording is, in principle, a relatively easy problem to deal with because there are accepted ways to "do it right." Much more complicated are cases in which seemingly arbitrary aspects of how a question is posed affect the values. Parducci (1974) has found that judged satisfaction with one's state in life may depend on the range of possible states mentioned in the question put to people. In an attempt to establish a dollar value for aesthetic degradation of the environment, Brookshire et al. (1976) asked visitors to Lake Powell how much they would be willing to pay in increased users' fees in order not to have an ugly (coal-fired) power plant looming on the opposite shore. They asked "Would you pay $1, $2, $3?" and so on, until the respondent answered "No" and then they retreated in decrements of a quarter (e.g., "Would you pay $5.75, $5.50, . . .?"). Rather different numerical values might have been obtained had the bidding procedure begun at $100 and decreased by steps of $10 or with other plausible variants. Any respondents who were not sure what they wanted in dollars and cents might naturally and necessarily look to the range of options presented, the difference between first and second options, and so on, for cues as to what are reasonable and plausible responses (Cummings et al., 1986; Smith and Desvousges, 1986).

At first glance, it might seem as though questions of value are the last redoubt of unaided intuition. Who knows better than an individual what he or she prefers? When people are considering simple, familiar events with which they have direct experience, it may be reasonable to assume that they have well-articulated opinions. Regarding the novel, global consequences potentially associated with CO_2-induced climatic change, nuclear meltdowns, or genetic engineering, that may not be the case. Our values may be incoherent, not thought through. In thinking about what are acceptable levels of risk, for example, we may be unfamiliar with the terms in which

issues are formulated (e.g., social discount rates, minuscule probabilities, or megadeaths). We may have contradictory values (e.g., a
strong aversion to catastrophic losses of life and a realization that we
are no more moved by a plane crash with 500 fatalities than by one
with 300). We may occupy different roles in life (parents, workers,
children) that produce clear-cut but inconsistent values. We may
vacillate between incompatible, but strongly held, positions (e.g.,
freedom of speech is inviolate, but should be denied to authoritarian
movements). We may not even know how to begin thinking about
some issues (e.g., the appropriate trade-off between the opportunity
to dye one's hair and a vague, minute increase in the probability of
cancer 20 years from now). Our views may undergo changes over
time (say, as we near the hour of decision or of experiencing the consequence) and we may not know which view should form the basis of
our decision.

An extreme, but not uncommon, situation is having no opinion
and not realizing it. In that state, we may respond with the first
thing that comes to mind once a question is asked and then com-
mit ourselves to maintaining that first expression and to mustering
support for it, while suppressing other views and uncertainties. As a
result, we may be stuck with stereotypical or associative responses,
generated without serious contemplation.

Once an issue has been evoked, it must be given a label. In a
world with few hard evaluative standards, such symbolic interpretations may be very important. While the facts of abortion remain
constant, individuals may vacillate in their attitude as they attach
and detach the label of murder. Figure II.5 shows two versions of the
same gamble, differing only in whether one consequence is labeled
a "sure loss" or an "insurance premium." Most people dislike the
former and like the latter. When these two versions are presented sequentially, people often reverse their preferences for the two options
(Hershey and Shoemaker, 1980). Figure II.6 shows a labeling effect
that produced a reversal of preference with practicing physicians;
most preferred treatment A over treatment B, and treatment D over
treatment C, despite the formal equivalence of A and C and of B and
D. Saving lives and losing lives afforded very different perspectives
on the same problem.

People solve problems, including the determination of their own
values, with what comes to mind. The more detailed, exacting, and
creative their inferential process, the more likely they are to think of
all they know about a question. The briefer that process becomes,

Insurance

Imagine that you must play a gamble in which you can lose but cannot win. Specifically, this gamble exposes you to:

1 chance in 4 to lose £200
(and 3 chances in 4 to lose nothing).

You can either take a chance with the gamble or insure against the £200 loss by buying a policy for a premium of £50. If you buy this insurance, you cannot lose £200, but you must pay the £50 premium.

Please indicate what you would do in this situation.

Preference

In this task you will be asked to choose between a certain loss and a gamble that exposes you to some chance of loss. Specifically, you must choose either:

Situation A: 1 chance in 4 to lose £200
                (and 3 chances in 4 to lose nothing)
or
Situation B: a certain loss of £50.

Of course, you would probably prefer not to be in either of these situations, but, if forced either to play the gamble (A) or to accept the certain loss (B), which would you prefer to do?

FIGURE II.5   Two formulations of a choice problem: insurance versus certain loss. SOURCE: Fischhoff et al., 1980.

the more they will be controlled by the relative accessibility of various considerations. Accessibility may be related to importance, but it is also related to the associations that are evoked, the order in which questions are posed, imaginability, concreteness, and other factors only loosely related to importance. As one example of how an elicitor may (perhaps inadvertently) control respondents' perspective, Turner (1980) observed a large difference in responses to a simple question such as "Are you happy?" on two simultaneous surveys of the same population (Figure II.7). The apparent source of the difference was that one (NORC) preceded the happiness question with a set of questions about married life. In the United States, married people are generally happier than unmarried people. Reminding them of that aspect of their life apparently changed the information that they brought to the happiness question.

It would be comforting to be able to say which way of phrasing these questions is most appropriate. However, there is no general answer. One needs to know why the question is being asked (Fischhoff

Lives Saved

Imagine that the U.S. is preparing for the outbreak of an unusual Asian disease, which is expected to kill 600 people. Two alternative programs to combat the disease have been proposed. The accepted scientific estimate of the consequences of the program are as follows:

If Program A is adopted, 200 people will be saved.

If Program B is adopted, there is 1/3 probability that 600 people will be saved, and 2/3 probability that no people will be saved.

Which of the two programs would you favor?

Lives Lost

If Program C is adopted, 400 people will die.

If Program D is adopted, there is 1/3 probability that nobody will die, and 2/3 probability that 600 people will die.

Which of the two programs would you favor?

FIGURE II.6   Two formulations of a choice problem: lives saved versus lives lost. SOURCE: Tversky and Kahneman, 1981. Copyright © 1981 by the American Association for the Advancement of Science.

and Furby, 1988). If one wants to predict the quality of casual encounters, then a superficial measure of happiness may suffice. However, an appraisal of national malaise or suicide potential may require a questioning procedure that evokes an appreciation of all components of respondents' lives. It has been known for some time that white interviewers evoke more moderate responses from blacks on race-related questions than do black interviewers. The usual response has been to match the races of interviewer and interviewee (Martin, 1980). This solution may be appropriate for predicting voting behavior or conversation in same-race bars, but not for predicting behavior of blacks in white-dominated workplaces.

The fact that one has a question is no guarantee that respondents have answers, or even that they have devoted any prior thought to the matter. When one must have an answer (say, because public input is statutorily required), there may be no substitute for an elicitation procedure that educates respondents about how they might look at the question. The possibilities for manipulation in such interviews are obvious. However, one cannot claim to be serving respondents' best interests (letting them speak their minds) by asking a question that only touches one facet of a complex and incompletely formulated set of views.

Refining Common Sense

Social scientists often find themselves in a no-win situation. If they describe their work in technical jargon, no one wants to listen. If they use plain language, no one feels a need to listen. Listeners feel that they "knew it all along" and that the social scientist was just "affirming the obvious" or "validating common sense." One possible antidote to this feeling is to point out the evidence showing that, in hindsight, people exaggerate how much they could have known in foresight, leading them to discount the informativeness of scientific

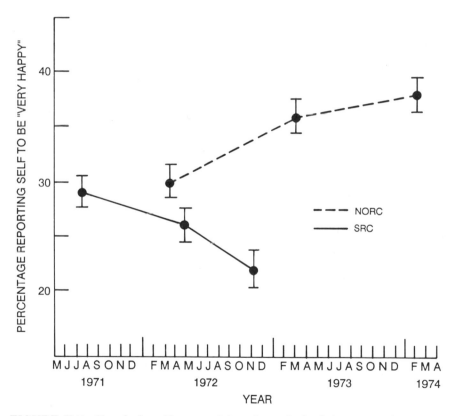

FIGURE II.7 Trends in self-reported happiness derived from sample surveys of the noninstitutionalized population of the continental United States aged 18 and over. Error bars demark ±1 standard error around sample estimate. SOURCE: Turner, 1980.

reports (Slovic and Fischhoff, 1977). A second antidote is to note
that common sense often makes contradictory predictions (e.g., two
heads are better than one versus too many cooks spoil the broth;
absence makes the heart grow fonder versus out of sight, out of
mind). Research is needed to determine which version of common
sense is correct or what their respective ranges of validity are. A
third strategy, adopted immediately below, is to present empirical
results that contradict conventional wisdom (Lazarsfeld, 1949).

Informing People About Risks

It is often claimed that people do not want to know very much
about the health risks they face, since such information makes them
anxious. Moreover, they cannot use that information very productively, even if it is given. If true, these claims would make it legitimate
for someone else (e.g., physicians, manufacturers, government) to decide what health (and therapeutic) risks are acceptable, and not to
invest too much effort on information programs. A number of investigators, however, have replaced anecdotal evidence with systematic
observation and have found that, by and large, people want to be
told about potential risks (Alfidi, 1971; Weinstein, 1980a). In clinical
settings, this desire has been observed with such risky practices as
psychotropic medication (Schwarz, 1978), endoscopy (Roling et al.,
1977), and oral contraceptives (Applied Management Sciences, 1978;
Joubert and Lasagna, 1975). Figure II.8 shows respondents' strong
opinions about the appropriate use of a pamphlet designed to explain the risks faced by temporary workers in a nuclear power plant.
Ninety percent of these individuals gave the most affirmative answer
possible to the question, "If you had taken such a job without being
shown this pamphlet, would you feel that you had been deprived of
necessary information?" (Fischhoff, 1981).

Risk-Taking Propensity

We all know that some people are risk takers and others are
risk avoiders; some are cautious, whereas others are rash. Indeed,
attitude toward risk might be one of the first attributes that comes
to mind when one is asked to describe someone else's personality. In
1962, Slovic compared the scores of 82 individuals on nine different
measures of risk taking. He found no consistency at all in people's
propensity for taking risks in the settings created by the various tests
(Slovic, 1962). Correlations ranged from −.35 to .34, with a mean of

When Should Pamphlet Be Shown?

	Definitely Yes					Definitely No
When they first sign up at the personnel office	X					
On the first morning when they first report to be driven out to job						X
On the morning when they arrive at the plant						X
Only when they ask for it explicitly						X
Not at all						X

FIGURE II.8   Opinions about the appropriate use of a pamphlet describing the risks associated with temporary work in a facility handling nuclear materials. Respondents were drawn from the readers of a student newspaper and from unemployed individuals at a state labor exchange. The "X" on each line represents the mean response to a question by the 173 individuals. SOURCE: Fischhoff, 1981.

.006. That is, people who are daring in one context may be timid in another, a result that has been replicated in numerous other studies (Davidshofer, 1976).

The surprising nature of these results may tell us something about ourselves as well as about the people we observe. One of the most robust psychological discoveries of the past 20 years has been identification of the *fundamental attribution error,* the tendency to view ourselves as highly sensitive to the demands of varying situations, but to see others as driven to consistent behavior by dominating personality traits (Nisbett and Ross, 1980). This misperception may be attributable to the fact that we typically see most others in only one role, as workers or spouses or parents or tennis players or drivers or whatever, in which the situational pressures are quite consistent. Thus, we may observe accurately the evidence available to us, but fail to understand the universe from which these data are drawn.

Protective Behavior

For years, the United States has been building flood control projects. Despite these great expenditures, flood losses today (in constant dollars) are greater than they were before this enterprise began. Apparently, the behavioral models of the dam and levee builders failed to account for the extent to which eliminating the recurrence of small-to-moderate floods reduced residents' (and particularly newcomers') sensitivity to flood dangers, which in turn led to overbuilding the flood plain. As a result, when the big floods come (about once every 100 years), exceeding the containment capacity of the protective structures, much more lies in their path (White, 1974).

The official response to this situation has been the National Flood Insurance Program (Kunreuther et al., 1978), designed according to economic models of human behavior, which assumes that flood plain residents are all-knowing, all-caring, and entirely "rational" (as defined by economics). Initially, premiums were greatly subsidized by the federal government to make the insurance highly attractive; these subsidies were to be withdrawn gradually once the insurance-buying habit was established. Unfortunately for the program, few people bought the insurance. The typical explanation for this failure was that residents expected the government to bail them out in the event of flood. However, a field survey found this speculation, too, to be in error. Flood plain residents reported that they expected no help, feeling that they were willingly bearing an acceptable risk. When residents thought about insurance at all, they seemed to rely on a melange of ad hoc principles like, "I can't worry about everything" and "The chances of getting a return (reimbursement) on my investment (premium) are too small," rather than on the concepts and procedures of economics (Kunreuther et al., 1978; Slovic et al., 1977).

ADHERENCE TO ESSENTIAL RULES OF SCIENCE

Looking hard at other sciences would reveal them to be similarly complicated, and similarly surprising. Sciences may not reveal their intricacies readily, but committed citizen activists have often proven themselves capable of mastering enough of the relevant science to be able to ask hard questions about risk issues that interest them (Figure II.4, for example, was created as a step toward this end). Many, of course, do not, and none could learn the hard questions about all of the sciences impinging on complex risk issues. This is, however, an option for those who care enough.

Short of such intense involvement, it is possible to ask some generic questions about almost any science. These are ways of asking "How good could it be?", given the conditions of its production.

Perhaps the most basic question that one can ask about any bit of science introduced into an environmental dispute, whether it be a single rodent bioassay or a full-blown risk analysis, is whether it actually represents a bit of science. In applied settings, one often finds evidence that fails to adhere to such essential rules of science as: (1) subjecting the study to critical peer review; (2) making all data available to other investigators; (3) evaluating the statistical reliability of results; (4) considering alternative explanations of the results; (5) relating new results to those already in the literature; and (6) pointing out critical assumptions that have not been empirically verified. Studies that fail to follow such procedures may be attempting to assume the rights, but not the responsibilities of science. Conversely, good science can come even from partisan sources (e.g., industry labs, environmental activists), if the rules are followed.

The definitiveness of science is bounded not only by the process by which it is conducted, but also by the object of its study. Some topics are simply easier than others, allowing for results clouded by relatively little uncertainty. Unfortunately for the rapid understanding and resolution of problems, risk management often demands understanding of inherently difficult topics.

This difficulty for risk managers can be seen as a by-product of one fortunate feature of the natural environment, namely, that the most fearsome events are quite infrequent. Major floods, disastrous plagues, and catastrophic tremors are all the exception rather than the rule. Social institutions attempt to constrain hazards of human origin so that the probability of their leading to disaster is low. However great their promised benefit, projects that might frequently kill large numbers of people are unlikely to be developed. The difficult cases are those in which the probability of a disaster is known to be low, but we do not know just how low. Unfortunately, quantitative assessment of very small probabilities is often very difficult (Fairley, 1977).

At times, one can identify a historical record that provides frequency estimates for an event related to the calamity in question. The U.S. Geological Survey has perhaps 75 years of reliable data on which to base assessments of the likelihood of large earthquakes (Burton et al., 1978). Iceland's copious observations of ice-pack movements over the last millennium provide a clue to the probability of an extremely cold year in the future (Ingram et al., 1978). The

absence of a full-scale meltdown in 500 to 1000 reactor-years of nuclear power plant operation sets some bounds on the probability of
future meltdowns (Weinberg, 1979). Of course, extrapolation from
any of these historical records is a matter of judgment. The great
depth and volume of artificial reservoirs may enhance the probability
of earthquakes in some areas. Increased carbon dioxide concentrations in the atmosphere may change the earth's climate in ways that
amplify or moderate yearly temperature fluctuations. Changes in design, staffing, and regulation may render the next 1000 reactor-years
appreciably different from their predecessors. Indeed, any attempt
to learn from experience and make a technology safer renders that
experience less relevant for predicting future performance.

Even when experts agree on the interpretation of records, a
sample of 1000 reactor-years or calendar-years may be insufficient.
If one believes the worst-case scenarios of some opponents of nuclear
power, a 0.0001 chance of a meltdown (per reactor-year) might seem
unconscionable. However, we will be into the next century before we
will have enough on-line experience to know with great confidence
whether the historical probability is really that low.

HOW DOES JUDGMENT AFFECT THE
RISK ESTIMATION PROCESS?

To the extent that historical records (or records of related systems) are unavailable, one must rely on conjecture. The more sophisticated conjectures are based on models such as the fault-tree
and event-tree analyses of a loss-of-coolant accident upon which the
Reactor Safety Study was based (U.S. Nuclear Regulatory Commission, 1975). As noted in Figure II.2, a fault tree consists of a logical
structuring of what would have to happen for an accident (e.g., a
meltdown) to occur. If sufficiently detailed, it will reach a level of
specificity for which one has direct experience (e.g., the operation
of individual valves). The overall probability of system failure is determined by combining the probabilities of the necessary component
failures.

The trustworthiness of such an analysis hinges on the experts'
ability to enumerate all major pathways to disaster and on the assumptions that underlie the modeling effort. Unfortunately, a modicum of systematic data and many anecdotal reports suggest that
experts may be prone to certain kinds of errors and omissions. Table

TABLE II.2 Some Problems in Structuring Risk Assessments

Failure to consider the ways in which human errors can affect technological systems.
   Example: Owing to inadequate training and control room design, operators at Three Mile Island repeatedly misdiagnosed the problems of the reactor and took inappropriate actions (Sheridan, 1980; U.S. Government, 1979).

Overconfidence in current scientific knowledge.
   Example: DDT came into widespread and uncontrolled use before scientists had even considered the possibility of the side effects that today make it look like a mixed, and irreversible, blessing (Dunlap, 1978).

Failure to appreciate how technological systems function as a whole.
   Example: The DC-10 failed in several early flights because its designers had not realized that decompression of the cargo compartment would destroy vital control systems (Hohenemser, 1975).

Slowness in detecting chronic, cumulative effects.
   Example: Although accidents to coal miners have long been recognized as one cost of operating fossil-fueled plants, the effects of acid rain on ecosystems were slow to be discovered (Rosencranz and Wetstone, 1980).

Failure to anticipate human response to safety measures.
   Example: The partial protection afforded by dams and levees gives people a false sense of security and promotes development of the flood plain. Thus, although floods are rarer, damage per flood is so much greater that the average yearly loss in dollars is larger than before the dams were built (Burton et al., 1978).

Failure to anticipate common-mode failures, which simultaneously afflict systems that are designed to be independent.
   Example: Because electrical cables controlling the multiple safety systems of the reactor at Browns Ferry, Alabama, were not spatially separated, all five emergency core-cooling systems were damaged by a single fire (Jennergren and Keeney, 1982; U.S. Government, 1975).

SOURCE: Fischhoff, Lichtenstein, et al., 1981a.

II.2 suggests some problems that might underlie the confident veneer of a formal model.

When the logical structure of a system cannot be described to allow computation of its failure probabilities (e.g., when there are large numbers of interacting systems), physical or computerized simulation models may be used. If one believes the inputs and the programmed interconnections, one should trust the results. What happens, however, when the results of a simulation are counterintuitive or politically awkward? There may be a strong temptation to try it again, adjusting the parameters or assumptions a bit, given that many of these are not known with certainty in the first place. Susceptibility to this temptation could lead to a systematic and subtle bias in modeling. At the extreme, models would be accepted only if they confirmed expectations.

Acknowledging the Role of Judgment

Although the substance of sciences differs greatly, sciences do have in common the fact that they are produced by the minds of mortals. Those minds may contain quite different facts, depending on the disciplines in which they were trained. However, it is reasonable to suppose that they operate according to similar principles when they are pressed to make speculations—taking them beyond the limits of hard data—in order to produce the sorts of assessments needed to guide risk managers.

Indeed, the need for judgment is a defining characteristic of risk assessment (*Federal Register* 49(100):21594–21661). Some judgment is, of course, a part of all science. However, the policy questions that hinge on the results of risk assessments typically demand greater scope and precision than can be provided by the "hard" knowledge that any scientific discipline currently possesses. As a result, risk assessors must fill the gaps as best they can. The judgments incorporated in risk assessments are typically those of esteemed technical experts, but they are judgments nonetheless, taking one beyond the realm of established fact and into the realm of educated opinions that cannot immediately be validated.

Judgment arises whenever materials scientists estimate the failure rates for valves subjected to novel conditions (Joksimovich, 1984; Östberg et al., 1977), whenever accident analysts attempt to recreate operators' perceptions of their situation prior to fatal mishaps (Kadlec, 1984; Pew et al., 1982), when toxicologists choose and weight extrapolation models (Rodricks and Tardiff, 1984; Tockman and Lilienfeld, 1984), when epidemiologists assess the reasons for nonresponse in a survey (Joksimovich, 1984; National Research Council, 1982), when pharmacokineticists consider how consumers alter the chemical composition of foods (e.g., by cooking and storage practices) before they consume them (National Research Council, 1983a; O'Flaherty, 1984), when physiologists assess the selection bias in the individuals who volunteer for their experiments (Hackney and

Linn, 1984; Rosenthal and Rosnow, 1969), when geologists consider how the construction of underground storage facilities might change the structure of the rock media and the flow of fluids through them (Sioshansi, 1983; Travis, 1984), and when psychologists wonder how the dynamics of a particular group of interacting experts affect the distribution of their responses (Brown, 1965; Davis, 1969; Hirokawa and Poole, 1986).

The process by which judgments are produced may be as varied as the topics they treat. Individual scientists may probe their own experience for clues to the missing facts. Reviewers may be sponsored to derive the best conclusions that the literature can provide. Panels of specialists may be convened to produce a collective best guess. Trained interviewers may use structured elicitation techniques to extract knowledge from others. The experts producing these judgments may be substantive experts in almost any area of science and engineering, risk assessment generalists who take it upon themselves to extrapolate from others' work, or laypeople who happen to know more than anyone else about particular facts (e.g., workers assessing how respirators are really used, civil defense officials predicting how evacuation plans will work).

Few experts would deny that they do not know all the answers. However, detailed treatments of the judgments they make in the absence of firm evidence are seldom forthcoming (*Federal Register* 49(100):21594–21661). There appear to be several possible causes for this neglect. Knowing which is at work in a particular risk assessment establishes what effect, if any, the informal treatment of judgment has had.

One common reason for treating the role of judgment lightly is the feeling that everyone knows that it is there, hence there is no point in repeating the obvious. Although this feeling is often justified, acting on it can have two deleterious consequences. One is that all consumers of an assessment may not share the same feeling. Some of these consumers may not realize that judgment is involved, whereas others may suspect that the judgments are being hidden for some ulterior purpose. The second problem is that failure to take this step precludes taking the subsequent steps of characterizing, improving, and evaluating the judgments involved.

A second, complementary reason for doing little about judgment is the belief that nothing much can be done, beyond a good-faith effort to think as hard as one can. Considering the cursory treatment of judgmental issues in most methodological primers for risk analysts, this perception is understandable. Considering the importance of doing something and the extensive research regarding what can be done, it is, however, not justifiable. Although the research is unfamiliar to most practicing analysts, the study and cultivation of judgment have proven tractable. The vulnerability of analyses to judgmental difficulties means that those who ignore judgment for this reason may miss a significant opportunity to perform at the state of the art.

A third reason for ignoring judgment is being rewarded for doing so. At times, analysts discern some strategic advantage to exaggerating the definitiveness of their work. At times, analysts feel that they must make a begrudging concession to the demands of political processes that attend only to those who speak with (unjustifiable) authority. At times, the neglect of judgment is (almost) a condition of employment, as when employers, hearings officials, or contracting agencies require statements of fact, not opinion.

Diagnosing the Role of Judgment

The first step in dealing with the judgmental aspects of risk assessments is identifying them. All risk assessment, and most contemporary science, can be construed as the construction of models. These include both procedures used to assess discrete hazards (e.g., accidents), such as probabilistic risk analysis, and procedures used to assess continuous hazards (e.g., toxicity), such as dose-response curves or structural-activity relationships. Although these models take many forms, all require a similar set of judgmental skills, which can be used as a framework for diagnosing where judgment enters into analyses (and, subsequently, how good it is and what can be done about it). These skills are:

1. *Identifying the active elements of the hazardous system being studied.* These may be the physical components of a nuclear power plant (e.g., the valves, controls, and piping) (U.S. Nuclear Regulatory Commission, 1983), the environmental factors affecting the dispersal of toxins from a waste disposal site (e.g., geologic structure, rainfall patterns, adjacent construction) (Pinder, 1984), or the potential predictors of cancer in an epidemiological study (Tockman and Lilienfeld, 1984).

2. *Characterizing the interrelationships among these elements.* Not everything is connected to everything else. Reducing the set of interconnections renders the model more tractable, its results

more comprehensible, and its data demands more manageable. The probabilistic risk analyst must judge which malfunctions in System X need to be considered when studying the performance of System Y. The epidemiologist needs to judge which interaction terms to include in regression models.

3. *Assessing the value of model parameters.* The amount of this kind of judgment varies greatly both across and within analyses. Some values have a sound statistical base (e.g., the number of chemical workers, as revealed by a decennial census), whereas others must be created from whole cloth (e.g., the sabotage rate at an as-yet-unconstructed plant 10 years in the future). Yet even the firmest statistics require some interpretation, for example, to correct for sampling and reporting biases or to adjust for subsequent changes in conditions.

4. *Evaluating the quality of the analysis.* Every analysis requires some summary statement of how good it is, whether for communicating its results to policymakers or for deciding whether to work on it more. Such evaluation requires consideration of both the substance and the purpose of the analysis. In both basic and applied sciences, the answer to "is the assessment good enough?" presupposes an answer to "good enough for what?"

5. *Adopting appropriate judgmental techniques.* Just as each stage in risk assessment requires different judgmental skills, it also requires different elicitation procedures. The reason for this is that each kind of information is organized in people's minds in a different way, and needs, therefore, to be extracted in a different way. For example, listing all possible mistakes that operators of a process-control industry might make is different than estimating how frequently each mistake will be made. The former requires heavy reliance on memory for instances of past errors, whereas the latter requires aggregation across diverse experiences and their extrapolation to future situations. Different experts (e.g., veteran operators, human factors theorists) may be more accustomed to thinking about the topic in one way rather than the other. Although transfer of information between these modes of thinking is possible, it may be far from trivial (Lachman et al., 1979; Tulving, 1972).

As noted earlier, studies with laypeople have found that seemingly subtle variations in how judgments are elicited can have large effects on the beliefs that are apparently revealed. These effects are most pronounced when people are least certain about how to respond, either because they do not know the answers or because they are unaccustomed to expressing themselves in the required terms. Thus, in extrapolating these results one must ask how expert the respondents are both in the topic requiring judgment and in using that response mode.

Assessing the Quality of the Judgment

If analysts have addressed the preceding steps conscientiously and left an audit trail of their work, all that remains is to review the protocol of the analysis to determine how heavily its conclusions depend on judgment and how adequate those judgments are likely to be. That evaluation should consider both the elicitation methods used and the judgmental capabilities of the experts. Ideally, the methods would have been empirically tested to show that they are: (1) compatible with the experts' mental representation of the problem, and (2) able to help the experts use their minds more effectively by overcoming common judgmental difficulties. Ideally, the experts would not only be knowledgeable about the topic, but also capable of translating that knowledge into the required judgments. The surest guarantees of that capability are having been trained in judgment or having provided judgments in conditions conducive to skill acquisition (e.g., prompt feedback).

How Good Are Expert Judgments?

As one might expect, considerably more is known about the judgmental processes of laypeople than about the judgmental processes of experts performing tasks in their areas of expertise. It is simply much easier to gain access to laypeople and create tasks about everyday events. Nonetheless, there are some studies of experts per se. In addition, there is some basis in psychological theory for extrapolating from the behavior of laypeople to that of experts. What follows is a selection of the kinds of problems that any of us may encounter when going beyond the available data, and which must be considered when weighing the usefulness of analyses estimating risks and benefits.

Sensitivity to Sample Size

Tversky and Kahneman (1971) found that even statistically sophisticated individuals have poor intuitions about the size of sample needed to test research hypotheses adequately. In particular, they expect small samples to represent the populations from which they were drawn to a degree that can only be assumed with much larger samples. This tendency leads them to gamble their research hypotheses on underpowered small samples, to place undue confidence in early data trends, and to underestimate the role of sampling variability in causing results to deviate from expectations (preferring instead to offer causal explanations for discrepancies). For example, in a survey of standard hematology texts, Berkson et al. (1939–1940) found that the maximum allowable difference between two successive blood counts was so small that it would normally be exceeded by chance 66 to 85 percent of the time. They mused about why instructors often reported that their best students had the most trouble attaining the desired standard.

Small samples mean low statistical power, that is, a small chance of detecting phenomena that really exist. Cohen (1962) surveyed published articles in a respected psychological journal and found very low power. Even under the charitable assumption that all underlying effects were large, a quarter of the studies had less than three chances in four of showing statistically significant results. He goes on to speculate that the one way to get a low-power study published is to keep doing it again and again (perhaps making subtle variations designed to "get it right next time") until a significant result occurs. Consequently, published studies may be unrepresentative of the set of conducted studies in a way that inflates the rate of spuriously significant results (beyond that implied by the officially reported "significance level"). Page (1981) has similarly shown the low power of representative toxicological studies. In designing such studies, one inevitably must make a trade-off between avoiding false alarms (e.g., erroneously calling a chemical a carcinogen) and misses (e.g., erroneously calling a chemical a noncarcinogen). Low power increases the miss rate and decreases the false alarm rate. Hence, wayward intuitions may lead to experimental designs that represent, perhaps inadvertently, a social policy that protects chemicals more than people.

Hindsight

Experimental work has shown that in hindsight people consistently exaggerate what could have been anticipated in foresight.

They tend not only to view what has happened as having been relatively inevitable, but also to view it as having appeared relatively inevitable before it happened. People believe that others should have been able to anticipate events much better than was actually the case. They even misremember their own predictions so as to exaggerate in hindsight what they knew in foresight (Fischhoff, 1980).

The revisionist history of strategic surprises (e.g., Lanir, 1982; Wohlstetter, 1962) argues that such misperceptions have vitiated the efforts of scholars and "scalpers" attempting to understand questions like, "Who goofed at Pearl Harbor?" These expert scrutinizers were not able to disregard the knowledge that they had only as a result of knowing how things turned out. Although it is flattering to believe that we personally would not have been surprised, failing to realize the difficulty of the task that faced the individuals about whom we are speculating may leave us very exposed to future surprises.

Methodological treatises for professional historians contain numerous warnings about related tendencies. One such tendency is telescoping the rate of historical processes, exaggerating the speed with which "inevitable" changes are consummated (Fischer, 1970). Mass immunization against poliomyelitis seems like such a natural idea that careful research is needed to show that its adoption met substantial snags, taking almost a decade to complete (Lawless, 1977). A second variant of hindsight bias may be seen in Barraclough's (1972) critique of the historiography of the ideological roots of Nazism; looking back from the Third Reich, one can trace its roots to the writings of many authors from whose writings one could not have projected Nazism. A third form of hindsight bias, also called "presentism," is to imagine that the participants in a historical situation were fully aware of its eventual importance ["Dear Diary, The Hundred Years' War started today" (Fischer, 1970)].

More directly relevant to the resolution of scientific disputes, Lakatos (1970) has argued that the "critical experiment," unequivocally resolving the conflict between two theories or establishing the validity of one, is typically an artifact of inappropriate reconstruction. In fact, "the crucial experiment is seen as crucial only decades later. Theories don't just give up, a few anomalies are always allowed. Indeed, it is very difficult to defeat a research programme supported by talented and imaginative scientists" (Lakatos, 1970:157–158).

Future generations may be puzzled by the persistence of the antinuclear movement after the 1973 Arab oil embargo guaranteed the future of nuclear power, or the persistence of nuclear advocates after Three Mile Island sealed the industry's fate—depending on how things turn out. Perhaps the best way to protect ourselves from the surprises and reprobation of the future in managing hazards is to "accept the fact of uncertainty and learn to live with it. Since no magic will provide certainty, our plans must work without it" (Wohlstetter, 1962:401).

Judging Probabilistic Processes

After seeing four successive heads in flips of a fair coin, most people expect a tails. Once diagnosed, this tendency is readily interpreted as a judgmental error. Commonly labeled the "gambler's fallacy" (Lindman and Edwards, 1961), it is one reflection of a strong psychological tendency to impose order on the results of random processes, making them appear interpretable and predictable (Kahneman and Tversky, 1972). Such illusions need not disappear with higher stakes or greater attention to detail. Feller (1968) offers one example in risk monitoring: Londoners during the Blitz devoted considerable effort to interpreting the pattern of German bombing, developing elaborate theories of where the Germans were aiming (and when to take cover). However, a careful statistical analysis revealed that the frequency distribution of bomb-hits in different sections of London was almost a perfect approximation of the Poisson (random) distribution. Dreman (1979) argues that the technical analysis of stock prices by market experts represents little more than opportunistic explication of chance fluctuations. Although such predictions generate an aura of knowing, they fail to outperform market averages.

Gilovich et al. (1985) found that, appearances to the contrary, basketball players have no more shooting streaks than one might expect from a random process generated by their overall shooting percentage. This result runs strongly counter to the conventional wisdom that players periodically have a "hot hand," attributable to specific causes like a half-time talk or dedication to an injured teammate. One of the few basketball experts to accept this result claimed that he could not act on it anyway. Fans would not forgive him if, in the closing minutes of a game, he had an inbound pass directed to a higher percentage shooter, rather than to a player with an apparent "hot hand" (even knowing that opposing players would cluster on that player, expecting the pass).

At times, even scientific enterprises seem to represent little more than sophisticated capitalization on chance. Chapman and Chapman (1969) found that clinical psychologists see patterns that they expect to find even in randomly generated data. O'Leary et al. (1974) observed that the theories of foreign affairs analysts are so complicated that any imaginable set of data can be interpreted as being consistent with them. Short of this extreme, it is generally true that, given a set of events (e.g., environmental calamities) and a sufficiently large set of possible explanatory variables (antecedent conditions), one can always devise a theory for retrospectively predicting the events to any desired level of proficiency. The price one pays for such overfitting is shrinkage, failure of the theory to work on a new sample of cases. The frequency and vehemence of warnings against such correlational overkill suggest that this bias is quite resistant to even extended professional training (Armstrong, 1975; Campbell, 1975; Crask and Parreault, 1977; Kunce et al., 1975).

Even when one is alert to such problems, it may be difficult to assess the degree to which one has capitalized on chance. For example, as a toxicologist, you are "certain" that exposure to chemical X is bad for one's health, so you compare workers who do and do not work with it in a particular plant for bladder cancer, but obtain no effect. So you try intestinal cancer, emphysema, dizziness, and so on, until you finally get a significant difference in skin cancer. Is that difference meaningful? Of course, the way to test these explanations or theories is by replication on new samples. That step, unfortunately, is seldom taken and is often not possible for technical or ethical reasons (Tukey, 1977).

A further unintuitive property of probabilistic events is regression to the mean, the tendency for extreme observations to be followed by less extreme ones. One depressing failure by experts to appreciate this fact is seen in Campbell and Erlebacher's (1970) article, "How regression artifacts in quasi-experimental evaluations can mistakenly make compensatory education look harmful" (because upon retest, the performance of the better students seems to have deteriorated). Similarly unfair tests may be created when one asks only if environmental management programs have, say, weakened strong industries or reduced productivity in the healthiest sectors of the economy.

Judging the Quality of Evidence

Since cognitive and evidential limits prevent scientists from providing all the answers, it is important to have an appraisal of how much they do know. It is not enough to claim that "these are the ranking experts in the field," for there are some fields in which the most knowledgeable individuals understand a relatively small portion of all there is to be known.

Weather forecasters offer some reason for encouragement (Murphy and Brown, 1983; Murphy and Winkler, 1984). There is at least some measurable precipitation on about 70 percent of the occasions for which they say there is a 70 percent chance of rain. The conditions under which forecasters work and train suggest the following prerequisites for good performance in probabilistic judgment:

- great amounts of practice;
- the availability of statistical data offering historical precipitation base rates (indeed, forecasters might be fairly well calibrated if they ignored the murmurings of their intuitions and always responded with the base rate);
- computer-generated predictions for each situation;
- a readily verifiable criterion event (measurable precipitation), offering clear feedback; and
- explicit admission of the imprecision of the trade and the need for training.

In experimental work, it has been found that large amounts of clearly characterized, accurate, and personalized feedback can improve the probability assessments of laypeople (e.g., Lichtenstein and Fischhoff, 1980).

Training professionals to assess and express their uncertainty is, however, a rarity. Indeed, the role of judgment is often acknowledged only obliquely. For example, civil engineers do not routinely assess the probability of failure for completed dams, even though approximately one dam in 300 collapses when first filled (U.S. Committee on Government Operations, 1978). The "Rasmussen" Reactor Safety Study (U.S. Nuclear Regulatory Commission, 1975) was an important step toward formalizing the role of risk in technological systems, although a subsequent review was needed to clarify the extent to which these estimates were but the product of fallible, educated judgment (U.S. Nuclear Regulatory Commission, 1978).

Ultimately, the quality of experts' assessments is a matter of judgment. Since expertise is so narrowly distributed, assessors are typically called upon to judge the quality of their own judgments. Unfortunately, an extensive body of research suggests that people are overconfident when making such assessments (Lichtenstein et al., 1982). A major source of such overconfidence seems to be failure to appreciate the nature and tenuousness of the assumptions on which judgments are based. To illustrate with a trivial example, when asked "To which country are potatoes native? (a) Ireland (b) Peru?", many people are very confident that answer (a) is true. The Irish potato and potato blight are familiar to most people; however, that is no guarantee of origin. Indeed, the fact that potatoes were not indigenous to Ireland may have increased their susceptibility to blight there.

Experts may be as prone to overconfidence as laypeople (in cases in which they, too, are pressed to evaluate judgments made regarding topics about which their knowledge is limited). For example, when several internationally known geotechnical experts were asked to predict the height of fill at which an embankment would fail and to give confidence intervals for their estimates, without exception, the true values fell outside the confidence intervals (Hynes and Vanmarcke, 1976), a result akin to that observed with other tasks and respondent populations (Lichtenstein et al., 1982). One of the intellectual challenges facing engineering is to systematize the role of judgment, both to improve its quality and to inform those who must rely on it in their decision making.

This basic pattern of results has proved so robust that it is hard to acquire much insight into the psychological processes producing it (Lichtenstein et al., 1982). One of the few effective manipulations is to force subjects to explain why their chosen answers might be wrong (Koriat et al., 1980). That simple instruction seems to prompt recall of contrary reasons that would not normally come to mind given people's natural thought processes, which seem to focus on retrieving reasons that support chosen answers. A second seemingly effective manipulation, mentioned earlier, is to train people intensively with personalized feedback that shows them how well they are calibrated.

Figures II.9 and II.10 show one sign of the limits that exist on the capacity of expertise and experience to improve judgment—in the absence of the conditions for learning enjoyed, say, by weather forecasters. Particle physicists' estimates of the value of several physical constants are bracketed by what might be called confidence intervals, showing the range of likely values within which the true

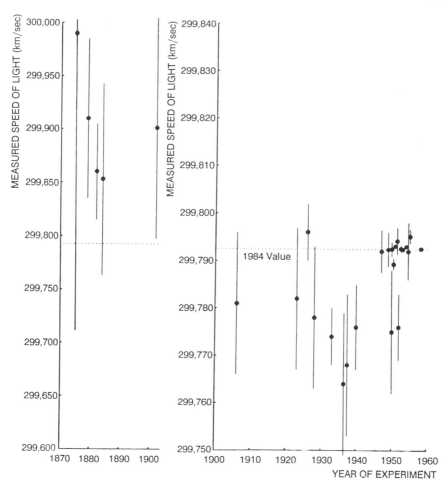

FIGURE II.9 Calibration of confidence in estimates of physical constants. SOURCE: Henrion and Fischhoff, 1986. Copyright © 1986 by the American Association of Physics Teachers.

value should fall, once it is known. Narrower intervals indicate greater confidence. These intervals have shrunk over time, as physicists' knowledge has increased. However, at most points, they seem to have been too narrow. Otherwise, the new best estimates would not have fallen so frequently outside the range of what previously seemed plausible. In an absolute sense, the level of knowledge represented here is extremely high and the successive best estimates lie extremely close to one another. However, the confidence intervals define what constitute surprises in terms of current physical theory. Unless the

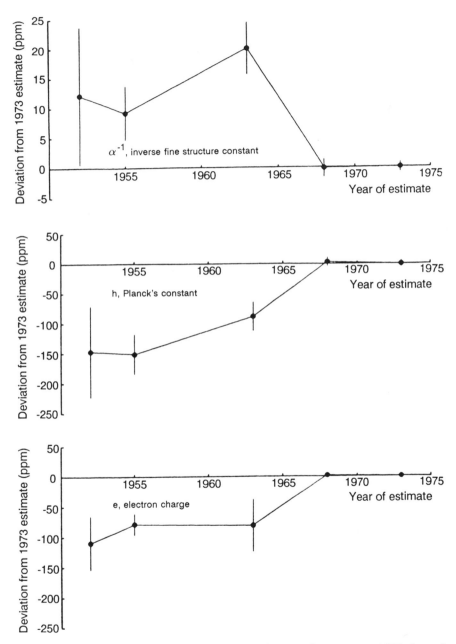

FIGURE II.10   Recommended values for fundamental constants, 1952 through 1973. SOURCE: Henrion and Fischhoff, 1986. Copyright © 1986 by the American Association of Physics Teachers.

possibility of overconfident judgment is considered, values falling outside the intervals suggest a weakness in theory.

SUMMARY

The science of risk provides a critical anchor for risk controversies. There is no substitute for that science. However, it is typically an imperfect guide. It can mislead if one violates any of a wide variety of intricate methodological requirements—including the need to use judgment judiciously (and to understand its limitations). The general nature of these assumptions was illustrated with examples drawn from the science of understanding human behavior. Sections IV through VI deal with the human anchors for risk controversies: the nature of their political tensions, the strategies that risk communicators can take in them, and psychological barriers to risk communication. The next section (III) deals with the interface between science and behavior, specifically ways in which science shapes and is shaped by the political process.

III

SCIENCE AND POLICY

SEPARATING FACTS AND VALUES

The first recommendation of the National Research Council's Committee on the Institutional Means for Assessment of Risks to Public Health (National Research Council, 1983b:7) was that:

> regulatory agencies take steps to establish and maintain a clear conceptual distinction between assessment of risks and considerations of risk management alternatives; that is, the scientific findings and policy judgments embodied in risk assessments should be explicitly distinguished from the political, economic, and technical considerations that influence the design and choice of regulatory strategies.

The principle of separating science and politics seems to be a cornerstone of professional risk management. Many of the antagonisms surrounding risk management seem due to the blurring of this distinction, resulting in situations in which science is rejected because it is seen as tainted by politics. As Hammond and Adelman (1976), Mazur et al. (1979), and others have argued, this distinction can help clear the air in debates about risk, which might otherwise fill up with half-truths, loaded language, and character assassinations. Even technical experts may fall prey to partisanship as they advance views on political topics beyond their fields of expertise, downplay facts they believe will worry the public, or make statements that cannot be verified.

Although a careful delineation between values and facts can help prevent values from hiding in facts' clothing, it cannot assure that a complete separation will ever be possible (Bazelon, 1979; Callen, 1976). The "facts" of a matter are only those deemed relevant to a particular problem, whose definition forecloses some action options and effectively prejudges others. Deciding what the problem is goes a long way to determining what the answer will be. Hence, the "objectivity" of the facts is always conditioned on the assumption that they are addressing the "right" problem, where "right" is defined in terms of society's best interest, not the interest of a particular party. The remainder of this section examines how our values determine what facts we produce and use, and how our facts shape our values.

Values Shape Facts

Without information, it may be hard to arouse concern about an issue, to allay fears, or to justify an action. But information is usually created only if someone has a use for it. That use may be pecuniary, scientific, or political. Thus, we may know something only if someone in a position to decide feels that it is worth knowing. Doern (1978) proposed that lack of interest in the fate of workers was responsible for the lack of research on the risks of uranium mining; Neyman (1979) wondered whether the special concern with radiation hazards had restricted the study of chemical carcinogens; Commoner (1979) accused oil interests of preventing the research that could establish solar power as an energy option. In some situations, knowledge is so specialized that all relevant experts may be in the employ of a technology's promoters, leaving no one competent to discover troublesome facts (Gamble, 1978). Conversely, if one looks hard enough for, say, adverse effects of a chemical, chance alone will produce an occasional positive finding. Although such spurious results are likely to vanish when studies are replicated, replications are the exception rather than the rule in many areas. Moreover, the concern raised by a faulty study may not be as readily erased from people's consciousness as from the scientific literature (Holden, 1980; Kolata, 1980; Peto, 1980). A shadow of doubt is hard to remove.

Legal requirements are an expression of society's values that may strongly affect its view of reality. Highway-safety legislation affects accident reports in ways that are independent of its effects on accident rates (V.L. Wilson, 1980). Crime-prevention programs may have similar effects, inflating the perceived problem by encouraging victims to report crimes (National Research Council, 1976). Although it is not always exploited for research purposes, an enormous legacy of medical tests has been created by the defensive medicine engendered by fear of malpractice. Legal concerns may also lead to the suppression of information, as doctors destroy "old" records that implicate them in the administration of diethylstilbestrol (DES) to pregnant women in the 1950s, employers fail to keep "unnecessary" records on occupational hazards, or innovators protect proprietary information (Lave, 1978; Pearce, 1979; Schneiderman, 1980).

Whereas individual scientists create data, it is the community of scientists and other interpreters who create facts by integrating data (Levine, 1974). Survival in this adversarial context is determined in part by what is right (i.e., truth) and in part by the staying power of those who collect particular data or want to believe in them. Scrutiny from both sides in a dispute is a valuable safeguard, likely to improve the quality of the analysis. Each side tries to eliminate erroneous material prejudicial to its position. If only one side scrutinizes, the resulting analyses will be unbalanced. Because staying with a problem requires resources, the winners in the marketplace of ideas may tend to be the winners in the political and economic marketplace.

Facts Shape Values

Values are acquired by rote (e.g., in Sunday school), by imitation, and by experience (Rokeach, 1973). The world we observe tells us what issues are worth worrying about, what desires are capable of fruition, and who we are in relation to our fellows. Insofar as that world is revealed to us through the prism of science, the facts it creates help shape our world outlook (R.P. Applebaum, 1977; Henshel, 1975; Markovic, 1970; Shroyer, 1970). The content of science's facts can make us feel like hedonistic consumers wrestling with our fellows, like passive servants of society's institutions, like beings at war with or at one with nature. The quantity of science's facts (and the coherence of their explication) may lower our self-esteem and enhance that of technical elites. The topics of science's inquiries may tell us that the important issues of life concern the mastery of others and of nature, or the building of humane relationships. Some argue that science can "anaesthetize moral feeling" (Tribe, 1972) by enticing us to think about the unthinkable. For example, setting an explicit value on human life in order to guide policy decisions may erode our social contract, even though we set such values implicitly by whatever decisions we make.

Even flawed science may shape our values. According to Wortman (1975), Westinghouse's poor evaluation of the Head Start program in the mid-1960s had a major corrosive effect on faith in social programs and liberal ideals. Weaver (1979) argued that whatever technical problems may be found with Inhaber's (1979) comparison of the risks of different energy sources, he succeeded in creating a new perspective that was deleterious to the opponents of nuclear power. As mentioned earlier, incorrect intuitions regarding the statistical power of statistical designs can lead to research that implicitly values chemicals more than people (Page, 1978, 1981). In designing such studies, one must make a trade-off between avoiding either false alarms (e.g., erroneously calling a chemical a carcinogen) or misses (e.g., not identifying a carcinogen as such). The decision to study many chemicals with relatively small samples both increases the miss rate and decreases the false-alarm rate. The value bias of such studies is compounded when scientific caution also becomes regulatory caution.

Where science concerns real-world objects, then the selection and characterization of those objects inevitably express attitudes toward them. Those attitudes may come from the risk managers who commission scientific studies, or they may come from the scientists who conduct them. In either case, the deepest link between science and politics may be in basic issues of definition. The next section discusses some of the subtle ways in which science can preempt or be captured by the policymaking process in its treatment of two basic concepts of risk management: risk and benefit.

MEASURING RISK

Which Hazards Are Being Considered?

The decision to decide whether a technology's risks are acceptable implies that, in the opinion of someone who matters, it may be too dangerous. Such issue identification is itself an action with potentially important consequences. Putting a technology on the decision-making agenda can materially change its fate by attracting attention to it and encouraging the neglect of other hazards. For example, concern about carbon-dioxide-induced climatic change (Schneider and Mesirow, 1976) changes the status of fossil fuels vis-à-vis nuclear power.

After an issue has been identified, the hazard in question must still be defined. Breadth of definition is particularly important. Are military and nonmilitary nuclear wastes to be lumped together in one broad category, or do they constitute separate hazards? Did the collision of two jumbo jets at Tenerife in the Canary Islands represent a unique miscommunication or a large class of pilot–controller impediments? Do all uses of asbestos make up a single industry or are brake linings, insulation, and so forth to be treated separately? Do hazardous wastes include residential sewage or only industrial solids (*Chemical and Engineering News*, 1980)? Grouping may convert a set of minor hazards into a major societal problem, or vice versa. Lead in the environment may seem worth worrying about, but lead solder in tuna fish cans may not. In recent years, isolated cases of child abuse have been aggregated in such a way that a persistent problem with a relatively stable rate of occurrence now appears as an epidemic demanding action.

Often the breadth of a hazard category becomes apparent only after the decision has been made and its implications experienced in practice. Some categories are broadened, for example, when precedent-setting decisions are applied to previously unrelated hazards. Other categories are narrowed over time as vested interests gain exceptions to the rules applying to the category in which their technology once belonged (Barber, 1979). In either case, different decisions might have been made had the hazard category been better defined in advance.

Definition of Risk

Managing technological risks has become a major topic in scientific, industrial, and public policy. It has spurred the development of some industries and prompted the demise of others. It has expanded the powers of some agencies and overwhelmed the capacity of others. It has enhanced the growth of some disciplines and changed the paths of others. It has generated political campaigns and counter-campaigns. The focal ingredient in all this has been concern over risk. Yet, the meaning of "risk" has always been fraught with confusion and controversy. Some of this conflict has been overt, as when a professional body argues about the proper measure of pollution or reliability for incorporation in a health or safety standard. More often, though, the controversy is unrecognized; the term risk is used in a particular way without extensive deliberations regarding the implications of alternative uses. Typically, that particular way follows custom in the scientific discipline initially concerned with the risk.

However, the definition of risk, like that of any other key term in policy issues, is inherently controversial. The choice of definition can affect the outcome of policy debates, the allocation of resources among safety measures, and the distribution of political power in society.

Dimensionality of Risk

The risks of a technology are seldom its only consequences. No one would produce it if it did not generate some benefits for someone. No one could produce it without incurring some costs. The difference between these benefits and nonrisk costs could be called the technology's net benefit. In addition, risk itself is seldom just a single consequence. A technology may be capable of causing fatalities in several ways (e.g., by explosions and chronic toxicity), as well as inducing various forms of morbidity. It can affect plants and animals as well as humans. An analysis of risk needs to specify which of these dimensions will be included. In general, definitions based on a single dimension will favor technologies that do their harm in a variety of ways (as opposed to those that create a lot of one kind of problem). Although it represents particular values (and leads to decisions consonant with those values), the specification of dimensionality (like any other specification) is often the inadvertent product of convention or other forces, such as jurisdictional boundaries (Fischhoff, 1984).

Summary Statistics

For each dimension selected as relevant, some quantitative summary is needed for expressing how much of that kind of risk is created by a technology. The controversial aspects of that choice can be seen by comparing the practices of different scientists. For some, the unit of choice is the annual death toll (e.g., Zentner, 1979); for others, deaths per person exposed or per hour of exposure (e.g., Starr, 1969); for others, it is the loss of life expectancy (e.g., Cohen and Lee, 1979; Reissland and Harries, 1979); for still others, lost working days (e.g., Inhaber, 1979). Crouch and Wilson (1982) have shown how the choice of unit can affect the relative riskiness of technologies. For example, today's coal mines are much less risky than those of 30 years ago in terms of accidental deaths per ton of coal, but marginally riskier in terms of accidental deaths per employee. The difference between measures is explained by increased productivity. The choice among measures is a policy question, with Crouch and Wilson suggesting that:

> From a national point of view, given that a certain amount of coal has to be obtained, deaths per million tons of coal is the more appropriate measure of risk, whereas from a labor leader's point of view, deaths per thousand persons employed may be more relevant (1982:13).

Other value questions may be seen in the units themselves. For example, loss of life expectancy places a premium on early deaths that is absent from measures treating all deaths equally; using it means ascribing particular worth to the lives of young people. Just counting fatalities expresses indifference to whether they come immediately after mishaps or following a substantial latency period (during which it may not be clear who will die). Whatever types of individuals are included in a category, they are treated as equals; the categories may include beneficiaries and nonbeneficiaries of a technology (reflecting an attitude toward that kind of equity), workers and members of the general public (reflecting an attitude toward that kind of voluntariness), or participants and nonparticipants in setting policy for the technology (reflecting an attitude toward that kind of voluntariness). Using the average of past casualties or the expectation of future fatalities means ignoring the distribution of risk over time; it treats technologies taking a steady annual toll in the same way as those that are typically benign, except for the rare catastrophic accident. When averages are inadequate, a case might be made for using one of the higher moments of the distribution of casualties over time or for incorporating a measure of the uncertainty surrounding estimates (Fischhoff, 1984).

Bounding the Technology

Willingness to count delayed fatalities means that a technology's effects are not being bounded in time (as they are, for example, in some legal proceedings that consider the time that passes between cause, effect, discovery, and reporting). Other bounds need to be set also, either implicitly or explicitly. One is the proportion of the fuel and materials cycles to be considered: To what extent should the risks be restricted to those people who enjoy the direct benefits of a technology or extended to cover those involved in the full range of activities necessary if those benefits are to be obtained? Crouch and Wilson (1982) offer an insightful discussion of some of these issues in the context of imported steel; the U.S. Nuclear Regulatory Commission (1983) has adopted a restrictive definition in setting safety goals for nuclear power (Fischhoff, 1983); much of the acrimony in the debates over the risks of competing energy technologies concerned treatment of the risks of back-up energy sources (Herbert et al., 1979; Inhaber, 1979). A second recurrent bounding problem is how far to go in considering higher-order consequences (i.e., when coping with one risk exposes people to another). As shown in Figure II.1, hazards begin with the human need the technology is designed to satisfy, and develop over time. One can look at the whole process or only at its conclusion. The more narrowly a hazard's moment in time is defined, the fewer the options that can be considered for managing its risks. A third issue of limits is how to treat a technology's partial contribution to consequences, for example, when it renders people susceptible to other problems or when it accentuates other effects through synergistic processes.

Concern

Events that threaten people's health and safety exact a toll even if they never happen. Concerns over accidents, illness, and unemployment occupy people even when they and their loved ones experience long, robust, and salaried lives. Although associated with risks, these consequences are virtual certainties. All those who know about them will respond to them in some way. In some cases, that response benefits the respondent, even if its source is an aversive event. For example, financial worries may prompt people to expand their personal skills or create socially useful innovations. Nonetheless, their resources have been diverted from other, perhaps preferred pursuits. Moreover, the accompanying stress can contribute to a variety of negative health effects, particularly when it is hard to control the threat (Elliot and Eisdorfer, 1982). Stress not only precipitates problems of its own, but can complicate other problems and divert the psychological resources needed to cope with them. Thus, concern about a risk may hasten the end of a marriage by giving the couple one more thing to fight about and that much less energy to look for solutions.

Hazardous technologies can evoke such concern even when they are functioning perfectly. Some of the response may be focused and purposeful, such as attempts to reduce the risk through personal and collective action. However, even that effort should be considered a cost of the technology because that time and energy might have been invested in something else (e.g., leisure, financial planning, improving professional skills) were it not for the technology. When many people are exposed to the risk (or are concerned about the exposure of their fellows), then the costs may be extensive. Concern may have even greater impact than the actual health and safety effects of the technology. Ironically, because the signs of stress are diffuse (e.g., a few more divorces, somewhat aggravated cardiovascular problems), it is quite possible for the size of the effects to be both intolerably large (considering the benefits) and undetectable (by current techniques).

Including concern among the consequences of a risky technology immediately raises two additional controversial issues. One centers on what constitutes an appropriate level of concern. It could be argued that concern should be proportionate to physical risk. There are, however, a variety of reasons why citizens might reasonably be concerned most about hazards that they themselves acknowledge to be relatively small (e.g., they feel that an important precedent is being set, that things will get worse if not checked, or that the chances for effective action are great) (see Section IV). The second issue is whether to hold a technology responsible for the concern evoked by people's perceptions of its risks or for the concern that would be evoked were people to share the best available technical knowledge. It is the former that determines actual concern; however, using it would mean penalizing some technologies for evoking unjustified concerns and rewarding others for having escaped the public eye.

MEASURING BENEFITS

Although the term risk management is commonly used for dealing with potentially hazardous technologies, few risk policies are concerned entirely with risk. Technologies would not be tolerated if they did not bring some benefit. Residual risk would not be tolerated if the benefits of additional reduction did not seem unduly expensive (to whoever is making the decision). As a result, some assessment of benefits is a part of all risk decisions, whether undertaken by institutions or by individuals. Faith in quantification makes formal cost–benefit analysis a part of many governmental decisions in the United States (Bentkover et al., 1985). However, a variety of procedures are possible, each with its own behavioral and ethical assumptions.

Definition of Benefit

Benefit assessment begins with a series of decisions that bound the analysis and specify its key terms. Together, these decisions provide an operational definition of what "benefit" means. Although they may seem technical and are often treated in passing, these decisions are the heart of an analysis. They express a social philosophy, elaborating what society holds to be important in a particular context. The ensuing analysis is "merely" an exercise in determining how well different policy options realize this philosophy. If the philosophy has not been interpreted, stated, and implemented appropriately, then the analysis becomes an exercise in futility.

The details of this definitional process in some ways parallel that for defining risk. Policymakers commission benefit assessments to help them make decisions; that is, to help them choose among alternative courses of action (including, typically, inaction). To make those decisions, they must (1) identify the policy alternatives (or options) that could be adopted; (2) circumscribe the set of policy-relevant consequences that these alternatives could create; (3) estimate the magnitude of each alternative's consequences were it adopted; (4) evaluate the benefits (and costs) that affected individuals would derive from these consequences; and (5) aggregate benefits across individuals. Defining the policymaking question is a precondition for commissioning any benefit assessment meant to serve it. For example, one cannot calculate the consequences of one particular policy without knowing the alternative policies that might come in its stead were it not adopted (and whose benefits would be foregone if it was). One cannot begin to assess and tally benefits without knowing which consequences and individuals fall within the agency's jurisdiction. Figure III.1 provides a summary of these definitional issues. Fischhoff and Cox (1985) discuss them in greater detail.

Once it has been determined what evaluations to seek, a method must be found for doing the seeking. There are two natural places to look for guidance regarding the evaluation of benefits: what people say and what people do. Methods relying on the former consider expressed preferences; methods relying on the latter consider revealed preferences. Each makes certain ethical and empirical assumptions regarding the nature of individual and societal behavior, the validity of which determines their applicability to particular situations (Driver et al., 1988).

Expressed Preferences

The most straightforward way to find out what people value, regarding safety or anything else, is to ask them. The asking can be done at the level of overall assessments (e.g., "Do you favor . . . ?"), statements of principle (e.g., "Should our society be risk averse regarding . . . ?"), or detailed trade-offs (e.g., "How much of a monetary sacrifice would you make in order to ensure . . . ?"). The vehicle for collecting these values could be public opinion polls (Conn, 1983), comments solicited at public hearings (Mazur, 1973; Nelkin, 1984), or detailed interviews conducted by decision analysts or counselors (Janis, 1982; Keeney, 1980). The advantages of these

IDENTIFYING THE SET OF POLICY OPTIONS
    Specifying details of each option
    Determining the range of variation
    Assessing the uncertainty surrounding implementation
    Anticipating the stability of the situation following inaction
    Determining the legitimacy of creating new options arising
        during the analysis

IDENTIFYING THE SET OF RELEVANT CONSEQUENCES
    Choosing consequences
        Scientific, legal, political, ethical grounds
        Public and private goods
    Specifying consequences
        Bounding in space
        Bounding in time
        Including higher-order consequences
        Including associated concern

ESTIMATING THE MAGNITUDES OF CONSEQUENCES
    Assessing the uncertainty around estimates
    Determining the risk assessor's attitude toward uncertainty
    Identifying deliberate bias in estimates
    Discerning the presuppositions in terms

EVALUATING BENEFITS FOR INDIVIDUALS
    Defining individuals
    Determining initial entitlements (willingness to pay versus
        willingness to accept)
    Identifying ultimate arbiter of benefit

AGGREGATING NET BENEFITS ACROSS INDIVIDUALS
    Looking for dominating alternatives (Pareto optimality)
    Exploring utilitarian solutions (potential Pareto improvements)
    Using group utility functions
    Resolving distributional inequities

FIGURE III.1   Steps in problem definition.   SOURCE: Fischhoff and Cox,
1985.

procedures are that they are current (in the sense of capturing today's
values), sensitive (in the sense of theoretically allowing people to say
whatever they want), specifiable (in the sense of allowing one to
ask the precise questions that interest policymakers), direct (in the
sense of looking at the preferences themselves and not how they
reveal themselves in application to some specific decision problem),
superficially simple (in the sense that you just ask people questions),
politically appealing (in the sense that they let "the people" speak),
and instructive (in the sense that they force people to think in a
focused manner about topics that they might otherwise ignore).

As discussed in Section II, however, a number of difficult conditions must be met if expressed preference procedures are to fulfill their promise. One is that the question asked must be the precise one needed for policymaking (e.g., "How much should you be paid in order to incur a 10 percent increase in your annual probability of an injury sufficiently severe to require at least one day of hospitalization, but not involving permanent disability?"), rather than an ill-defined one, such as "do you favor better roads?" or "is your job too risky?" (In response, a thoughtful interviewee might ask, "What alternatives should I be considering? Am I allowed to consider who pays for improvements?") One response to the threat of ambiguity is to lay out all details of the evaluation question to respondents (Fischhoff and Furby, 1988). A threat to this solution is that the full specification will be so complex and unfamiliar as to pose an overwhelming inferential task. To avoid the incompletely considered, and potentially labile, responses that might arise, one must either adjust the questions to the respondents or the respondents to the questions. The former requires an empirically grounded understanding of what issues people have considered and how they have thought about them. This understanding allows one to focus the interview on the areas in which people have articulated beliefs, to provide needed elaborations, and to avoid repeating details that correspond to respondents' default assumptions (and could, therefore, go without saying).

If the gap between policymakers' questions and respondents' answers is too great to be bridged in a standard interviewing session, then it may be necessary either to simplify the questions or to complicate the session. A structured form of simplification is offered by techniques, such as multi-attribute utility theory, which decompose complex questions into more manageable components, each of which considers a subsidiary evaluation issue (Keeney and Raiffa, 1976). The structuring of these questions allows their recomposition into overall evaluations, which are interpreted as representing the summary judgments that respondents would have produced if they had unlimited mental computational capacity. The price paid for this potential simplification is the need to answer large numbers of simple, formal, and precise questions.

Where it becomes impossible to bring the question "down" to the level of the respondent, there still may be some opportunity to bring the respondent "up" to the level of the question. Ways of enabling respondents to realize their latent capability for thinking meaningfully about questions include talking with them about the issues, including them in focused group discussions, suggesting alternative perspectives (for their consideration), and giving them time to ruminate over their answers.

Revealed Preferences

The alternative to words is action. This collection of techniques assumes that people's overt actions can be interpreted to reveal the preferences that motivated them. The great attraction of such procedures is that they are based on real acts, whose consequences are presumably weightier than those of even the most intelligently conducted interview. They focus on possibilities, rather than just desires.

By concentrating on current, real decisions, these procedures are also strongly anchored in the status quo. It is today's work, with today's constraints, that conditions the behavior observed. If today's society inhibits people's ability to act in ways that express their fundamental values, then revealed preference procedures lose their credibility (whereas expressed preferences, at least in principle, allow people to raise themselves above today's reality). Thus, if one feels that advertising, or regulation, or monopoly pressures have distorted contemporary evaluations of some products or consequences, then revealing those values does not yield a guide to true worth. Relying on those values for policymaking would mean enshrining today's imperfections (and inequities) in tomorrow's world.

The commitment to observing actual behavior also makes these procedures particularly vulnerable to deviations from optimality. A much smaller set of inferences separates people's true values from their expressed preferences than from their overt behavior. On the one hand, this means that people must complete an even more complex series of inferences in order to do what they want than to say what they want. On the other hand, investigators must make even more assumptions in order to infer underlying values from what they observe. Thus, for example, it is difficult enough to determine how much compensation one would demand to accept an additional injury risk of magnitude X in one's job. Implementing that policy in an actual decision also requires that suitable options be available and that their consequences be accurately perceived. If those conditions of informed consent are not met, then the interpretation of pay–danger relationships may be quite tenuous. Workers may be coercing their employer into compensating them for imagined risks;

or, they may be coerced into accepting minimal compensation by an employer cognizant of a depressed job market.

The most common kind of revealed preference analysis is also the most common kind of economic analysis: interpreting marketplace prices as indicating the true value of goods. If the goods whose values are of interest (e.g., health risks) are not traded directly, then a value may be inferred by conceptualizing the goods that are traded (e.g., jobs) as representing a bundle of consequences (e.g., risks, wages, status). Analytic techniques may then be used to discern the price that markets assign to each consequence individually, by looking at its role in determining the price paid for various goods that include it.

These regression-based procedures rest on a well-developed theoretical foundation describing why (under conditions of a free market, optimal decision making, and informed consent) prices should reveal the values that people ascribe to things (Bentkover et al., 1985). The same general thought has been applied heuristically in various schemes designed to discern the values revealed in decisions (ostensibly) taken by society as a whole or by individuals under less constrained conditions. These analyses include attempts to see what benefits society demands for tolerating the risks of different technologies (Starr, 1969), what risks people seem to accept in their everyday lives (B. Cohen and Lee, 1979; R. Wilson, 1979), and what levels of technological risk escape further regulation (Fischhoff, 1983; U.S. Nuclear Regulatory Commission, 1982). These attempts are typically quite ad hoc, with no detailed methodology specifying how they should be conducted. The implicit underlying theory assumes, in effect, that whatever is, is right and that present arrangements are an appropriate basis for future policies. Thus, these procedures can guide future decisions only if one believes that society as a whole currently gets what it wants, even with regard to regulated industries, unregulated semimonopolies, and poorly understood new technologies. Extracting useful information from them requires a very detailed assessment of the procedures that they use, the existing reality that they endorse, and the kinds of behavior that they study.

Ascertaining the validity of the theory underlying approaches to measuring "benefit" that assume optimality has often proven difficult, for what can best be described as philosophical reasons. Some investigators find it implausible that people do anything other than optimize their own best interest when making decisions, maintaining that society would not be functioning so well were it not for this ability. These investigators see their role as discerning what people are trying to optimize (i.e., what values they ascribe to various consequences).

The contrary position argues that this belief in optimality is tautological, in that one can always find something that people could be construed as trying to optimize. Looking at how decisions are actually made shows that they are threatened by all the problems that can afflict expressed preferences. Thus, for example, consumers may make suboptimal choices because a good is marketed in a way that evokes only a portion of their values, or because they unwittingly exaggerate their ability to control its risks (Svenson, 1981; Weinstein, 1980a).

Because of the philosophical differences between these positions, relatively little is known about the general sensitivity of conclusions drawn from analyses that assume optimality to deviations from optimality. The consumer of such analyses is left to discern how far conditions deviate from optimal decision making by informed individuals in an unconstrained marketplace and, then, how far those deviations threaten the conclusions of the analyses.

SUMMARY

Science is a product of society; as such, it reflects the values of its creators. That reflection may be deliberate, as when young people decide how to dedicate their lives and research institutes decide how to stay solvent. Or, it may be unconscious, as scientists routinely apply value-laden procedures and definitions just because that was what they learned to do in school. Conversely, society is partly a product of science. That influence may be direct, as when science shapes the conditions under which people live (e.g., how prosperous they are, what industries confront them). Or it may be indirect, as when science defines our relationship with nature or raises specific fears. Understanding these interdependencies is essential to, on the one hand, discerning the objective content versus inherently subjective science and, on the other hand, directing science to serve socially desired ends. An understanding of these relationships is also necessary to appropriately interpret the conflicts between lay and expert opinions that constitute the visible core of many risk controversies. The diagnoses of these conflicts are discussed in Section IV.

IV

THE NATURE OF THE
CONTROVERSY

A public opinion survey (Harris, 1980) reported the following three results:

1. Among four "leadership groups" (top corporate executives, investors and lenders, congressional representatives, and federal regulators), 94 to 98 percent of all respondents agreed with the statement "even in areas in which the actual level of risk may have decreased in the past 20 years, our society is significantly more aware of risk."

2. Between 87 and 91 percent of those four leadership groups felt that "the mood of the country regarding risk" will have a substantial or moderate impact "on investment decisions—that is, the allocation of capital in our society in the decade ahead." (The remainder believed that it would have a minimal impact, no impact at all, or were not sure.)

3. No such consensus was found, however, when these groups were asked about the appropriateness of this concern about risk. A majority of the top corporate executives and a plurality of lenders believed that "American society is overly sensitive to risk," whereas a large majority of congressional representatives and federal regulators believed that "we are becoming more aware of risk and taking realistic precautions." A sample of the public endorsed the latter statement over the former by 78 to 15 percent.

In summary, there is great agreement that risk decisions will have a major role in shaping our society's future and that those decisions will, in turn, be shaped by public perceptions of risk. There is, however, much disagreement about the appropriateness of those perceptions. Some believe the public to be wise; others do not. These contrary beliefs imply rather different roles for public involvement in risk management. As a result, the way in which this disagreement is resolved will affect not only the fate of particular technologies, but also the fate of our society and its social organization.

To that end, various investigators have been studying how and how well people think about risks. Although the results of that research are not definitive as yet, they do clearly indicate that a careful diagnosis is needed whenever the public and the experts appear to disagree. It is seldom adequate to attribute all such discrepancies to public misperceptions of the science involved. From a factual perspective, that assumption is often wrong; from a societal perspective, it is generally corrosive by encouraging disrespect among the parties involved. When the available research data do not allow one to make a confident alternative diagnosis, a sounder assumption is that there is some method in the other party's apparent madness. This section offers some ways to find that method. Specifically, it offers six reasons why disagreements between the public and the experts need not be interpreted merely as clashes between actual and perceived risks.

THE DISTINCTION BETWEEN "ACTUAL" AND "PERCEIVED" RISKS IS MISCONCEIVED

Although there are actual risks, nobody knows what they are. All that anyone does know about risks can be classified as perceptions. Those assertions that are typically called actual risks (or facts or objective information) inevitably contain some element of judgment on the part of the scientists who produce them. In this light, what is commonly called the conflict between actual and perceived risk is better thought of as the conflict between two sets of risk perceptions: those of ranking scientists performing within their field of expertise and those of anybody else. The element of judgment is most minimal when all the experts do is to assess the competence of a particular study conducted within an established paradigm. It grows with the degree to which experts must integrate results from diverse studies or extrapolate from a domain in which results are readily obtainable to another in which they are really needed (e.g., from animal studies to human effects). Judgment becomes all when there are no (credible) available data, yet a policy decision requires some assessment of a particular fact. Section II discusses at length the trustworthiness of such judgments.

The expert opinions that make up the scientific literature aspire to be objective in two senses, neither of which can ever be achieved absolutely and neither of which is the exclusive province of technical experts. One meaning of objectivity is reproducibility: one expert should be able to repeat another's study, review another's protocol, reanalyze another's data, or recap another's literature summary and reach the same conclusions about the size of an effect. Clearly, as the role of judgment increases in any of these operations, the results become increasingly subjective. Typically, reproducibility should decrease (and subjectivity increase) to the extent that a problem attracts scientists with diverse training or falls into a field that has yet to reach consensus on basic issues of methodology.

The second sense of objectivity means immune to the influence by value considerations. One's interpretations of data should not be biased by one's political views or pecuniary interests. Applied sciences naturally have developed great sensitivity to such problems and are able to invoke some penalties for detected violations. There is, however, little possibility of regulating the ways in which values influence other acts, such as one's choice of topics to study or ignore. Some of these choices might be socially sanctioned, in the sense that one's values are widely shared (e.g., deciding to study cancer because it is an important problem); other choices might be more personal (e.g., not studying an issue because one's employer does not wish to have troublesome data created on that topic). Although a commitment to separating issues of fact from issues of value is a fundamental aspect of intellectual hygiene, a complete separation is never possible (see Section III).

At times, this separation is not even desired—as when experts offer their views on how risks should be managed. Because they mix questions of fact and value, such views might be better thought of as the opinions of experts rather than as expert opinions, a term that should be reserved for expressions of substantive expertise. It would seem as though members of the public are the experts when it comes to striking the appropriate trade-offs between costs, risks, and benefits. That expertise is best tapped by surveys, hearings, and political campaigns.

Of course, there is no all-purpose public any more than there are all-purpose experts. The ideal expert on a matter of fact has studied that particular issue and is capable of rendering a properly qualified opinion in a form useful to decision makers. Using the same criteria for selecting value experts might lead one to philosophers, politicians, psychologists, sociologists, clergy, intervenors, pundits, shareholders, or well-selected bystanders. Thus, one might ask, "in what sense," whenever someone says "expert" or "public" (Schnaiburg, 1980; Thompson, 1980). This appendix uses "expert" in the restrictive sense and "public" or "laypeople" to refer to everyone else, including scientists in their private lives.

LAYPEOPLE AND EXPERTS ARE SPEAKING
DIFFERENT LANGUAGES

Explicit risk analyses are a fairly new addition to the repertoire of intellectual enterprises. As a result, risk experts are only beginning to reach consensus on basic issues of terminology and methodology, such as how to define risk (see Section III). Their communications to the public reflect this instability. They are only beginning to express a sufficiently coherent perspective to help the public sort out the variety of meanings that "risk" could have. Under these circumstances some miscommunication may be inevitable. Studies (Slovic et al., 1979, 1980) have found that when expert risk assessors are asked to assess the risk of a technology on an undefined scale, they tend to respond with numbers that approximate the number of recorded or estimated fatalities in a typical year. When asked to estimate average year fatalities, laypeople produce fairly similar numbers. When asked to assess risk, however, laypeople produce quite different responses. These estimates seem to be an amalgam of their average-year fatality judgments, along with their appraisal of other features, such as a technology's catastrophic potential or how equitably its risks are distributed. These catastrophic potential judgments match those of the experts in some cases, but differ in others (e.g., nuclear power).

On semantic grounds, words can mean whatever a population group wants them to mean, as long as that usage is consistent and does not obscure important substantive differences. On policy grounds, the choice of a definition is a political question regarding what a society should be concerned about when dealing with risk. Whether we attach special importance to potential catastrophic losses of life or convert such losses to expected annual fatalities (i.e., multiply the potential loss by its annual probability of occurrence) and add them to the routine toll is a value question—as would be a decision to weight those routine losses equally rather than giving added weight to losses among the young (or among the nonbeneficiaries of a technology).

For other concepts that recur in risk discussions, the question of what they do or should mean is considerably murkier. It is often argued, for example, that different standards of stringency should apply to voluntarily and involuntarily incurred risks (e.g., Starr, 1969). Hence, for example, skiing could (or should) legitimately be a more hazardous enterprise than living below a major dam. Although there is general agreement among experts and laypeople about the voluntariness of food preservatives and skiing, other technologies are more problematic (Fischhoff et al., 1978b; Slovic et al., 1980). There is considerable disagreement within expert and lay groups in their ratings of the voluntariness of technologies such as prescription antibiotics, commercial aviation, handguns, and home appliances. These disagreements may reflect differences in the exposures considered; for example, use of commercial aviation may be voluntary for vacationers, but involuntary for certain business people (and scientists). Or, they may reflect disagreements about the nature of society or the meaning of the term. For example, each decision to ride in a car may be voluntarily undertaken and may, in principle, be foregone (i.e., by not traveling or by using an alternative mode of transportation); but in a modern industrial society, these alternatives may be somewhat fictitious. Indeed, in some social sets, skiing may be somewhat involuntary. Even if one makes a clearly volitional decision, some of the risks that one assumes may be indirectly and involuntarily imposed on one's family or the society that must pick up the pieces (e.g., pay for hospitalization due to skiing accidents).

Such definitional problems are not restricted to "social" terms such as "voluntary." Even a technical term such as "exposure" may be consensually defined for some hazards (e.g., medical x rays), but not for others (e.g., handguns). In such cases, the disagreements within expert and lay groups may be as large as those between them. For orderly debate to be possible, one needs some generally accepted definition for each important term—or at least a good translating dictionary. For debate to be useful, one needs an explicit analysis of whether each concept, so defined, makes a sensible basis for policy. Once they have been repeated often enough, ideas such as the importance of voluntariness or catastrophic potential tend to assume a life of their own. It does not go without saying that society should set a double standard on the basis of voluntariness or catastrophic potential, however they are defined.

LAYPEOPLE AND EXPERTS ARE SOLVING DIFFERENT PROBLEMS

Many debates turn on whether the risk associated with a particular configuration of a technology is acceptable. Although these disagreements may be interpreted as reflecting conflicting social values or confused individual values, closer examination suggests that the acceptable-risk question itself may be poorly formulated (Otway and von Winterfeldt, 1982).

To be precise, one does not accept risks—one accepts options that entail some level of risk among their consequences. Whenever the decision-making process has considered benefits or other (nonrisk) costs, the most acceptable option need not be the one with the least risk. Indeed, one might choose (or accept) the option with the highest risk if it had enough compensating benefits. The attractiveness of an option depends on its full set of relevant positive and negative consequences (Fischhoff, Lichtenstein, et al., 1981).

In this light, the term "acceptable risk" is ill defined unless the options and consequences to be considered are specified. Once the options and consequences are specified, "acceptable risk" might be used to denote the risk associated with the most acceptable alternative. When using that designation, it is important to remember its context dependence. That is, people may disagree about the acceptability of risks not only because they disagree about what those consequences are (i.e., they have different risk estimates) or because they disagree about how to evaluate the consequences (i.e., they have different values), but also because they disagree about what consequences and options should be considered.

Some familiar policy debates might be speculatively attributed, at least in part, to differing conceptions of what the set of possible options is. For example, saccharin (with its risks) may look unacceptable when compared with life without artificial sweeteners (one possible alternative option). Artificial sweeteners may, however, seem more palatable when the only alternative option considered is another sweetener that appears to be more costly and more risky. Or, nuclear power may seem acceptable when compared with alternative sources of generating electricity (with their risks and costs), but not so acceptable when aggressive conservation is added to the option set. Technical people from the nuclear industry seem to prefer the narrower problem definition, perhaps because they prefer to concentrate on the kinds of solutions most within their domain of expertise. Citizens involved in energy debates may feel themselves less narrowly bound; they may also be more comfortable with solutions, such as conservation, that require their kind of expertise (Bickerstaffe and Peace, 1980).

People who agree about the facts and share common values may still disagree about the acceptability of a technology because they have different notions about which of those values are relevant to a particular decision. For example, all parties may think that equity is a good thing in general, without agreeing also that energy policy is the proper arena for resolving inequities. For example, some may feel that both those new inequities caused by a technology and those old ones endemic to a society are best handled separately (e.g., through the courts or with income policies).

Thus, when laypeople and experts disagree about the acceptability of a risk, one must always consider the possibility that they are addressing different problems, with different sets of alternatives or different sets of relevant consequences. Assuming that each group has a full understanding of the implications of its favored problem definition, the choice among definitions is a political question. Unless a forum is provided for debating problem definitions, these concerns may emerge in more indirect ways (Stallen, 1980).

DEBATES OVER SUBSTANCE MAY DISGUISE BATTLES OVER FORM, AND VICE VERSA

In most political arenas, the conclusion of one battle often sets some of the initial conditions for its successor. Insofar as risk management decisions are shaping the economic and political future of a country, they are too important to be left to risk managers (Wynne, 1980). When people from outside the risk community enter risk battles, they may try to master the technical details or they may concentrate on monitoring and shaping the risk management process itself. The latter strategy may exploit their political expertise and keep them from being outclassed on technical issues. As a result, their concern about the magnitude of a risk may emerge in the form of carping about how it has been studied. They may be quick to criticize any risk assessment that does not have such features as eager peer review, ready acknowledgment of uncertainty, or easily accessible documentation. Even if they admit that these features are consonant with good research, scientists may resent being told by laypeople how to conduct their business even more than they resent being told by novices what various risks really are.

Lay activists' critiques of the risk assessment process may be no less irritating, but somewhat less readily ignored, when they focus on the way in which scientists' agendas are set. As veteran protagonists in hazard management struggles know, without scientific information it may be hard to arouse and sustain concern about an issue, to allay inappropriate fears, or to achieve enough certainty to justify action.

However, information is, by and large, created only if someone has a (professional, political, or economic) use for it. Whether the cause is fads or finances, failure to study particular topics can thwart particular parties and may lead them to impugn the scientific process.

At the other extreme, debates about political processes may underlie disputes that are ostensibly about scientific facts. As mentioned earlier, the definition of an acceptable-risk problem circumscribes the set of relevant facts, consequences, and options. This agenda setting is often so powerful that a decision has effectively been made once the definition is set. Indeed, the official definition of a problem may preclude advancing one's point of view in a balanced fashion. Consider, for example, an individual who is opposed to increased energy consumption but is asked only about which energy source to adopt. The answers to these narrower questions provide a de facto answer to the broader question of growth. Such an individual may have little choice but to fight dirty, engaging in unconstructive criticism, poking holes in analyses supporting other positions, or ridiculing opponents who adhere to the more narrow definition. This apparently irrational behavior can be attributed to the rational pursuit of officially unreasonable objectives.

Another source of deliberately unreasonable behavior arises when participants in technology debates are in it for the fight. Many approaches to determining acceptable-risk levels (e.g., cost–benefit analyses) make the political-ideological assumption that our society is sufficiently cohesive and common-goaled that its problems can be resolved by reason and without struggle. Although such a "get on with business" orientation will be pleasing to many, it will not satisfy all. For those who do not believe that society is in a fine-tuning stage, a technique that fails to mobilize public consciousness and involvement has little to recommend it. Their strategy may involve a calculated attack on what they interpret as narrowly defined rationality (Campen, 1985).

A variant on this theme occurs when participants will accept any process as long as it does not lead to a decision. Delay, per se, may be the goal of those who wish to preserve some status quo. These may be environmentalists who do not want a project to be begun or industrialists who do not want to be regulated. An effective way of thwarting practical decisions is to insist on the highest standards of scientific rigor.

LAYPEOPLE AND EXPERTS DISAGREE
ABOUT WHAT IS FEASIBLE

Laypeople are often berated for misdirecting their efforts when they choose risk issues on which to focus their energies. However, a more careful diagnosis can often suggest several defensible strategies for setting priorities. For example, Zentner (1979) criticizes the public because its rate of concern about cancer (as measured by newspaper coverage) is increasing faster than the cancer rate. One reasonable explanation for this pattern is that people may believe that too little concern has been given to cancer in the past (e.g., our concern for acute hazards like traffic safety and infectious disease allowed cancer to creep up on us). A second is that people may realize that some forms of cancer are among the only major causes of death that experience increasing rates.

Systematic observation and questioning are, of course, needed to tell whether these speculations are accurate (and whether the assumption of rationality holds in this particular case). False positives in divining people's underlying rationality can be as deleterious as false negatives. Erroneously assuming that laypeople understand an issue may deny them a needed education; erroneously assuming that they do not understand may deny them a needed hearing. Pending systematic studies, these error rates are likely to be determined largely by the rationalist or emotionalist cast of one's view of human nature.

Without solid evidence to the contrary, perhaps the most reasonable general assumption is that people's investment in problems depends on their feelings of personal efficacy. That is, they are unlikely to get involved unless they feel that they can make a difference, personally or collectively. In this light, their decision-making process depends on a concern that is known to influence other psychological processes: perceived feelings of control (Seligman, 1975). As a result, people will deliberately ignore major problems if they see no possibility of effective action. Here are some reasons why they might reject a charge of "misplaced priorities" when they neglect a hazard that poses a large risk:

- the hazard is needed and has no substitutes;
- the hazard is needed and has only riskier substitutes;
- no feasible scientific study can yield a sufficiently clear and incontrovertible signal to legitimate action;

- the hazard is distributed naturally, and hence cannot be controlled;
- no one else is worried about the risk in question, and thus no one will heed messages of danger or be relieved by evidence of safety; and
- no one is empowered to or able to act on the basis of evidence about the risk.

Thus, the problems that actively concern people need not be those whose resolution they feel should rank highest on society's priorities. For example, one may acknowledge that the expected deaths from automobile accidents over the next century are far greater than those expected from nuclear power, and yet still be active only in fighting nuclear power out of the conviction, "Here, I can make a difference. This industry is on the ropes now. It's important to move in for the kill before it becomes as indispensable to American society as automobile transportation."

Thus, differing priorities between experts and laypeople may not reflect disagreements about the size of risks, but differing opinions on what can be done about them. At times, the technical knowledge or can-do perspective of the experts may lead them to see a broader range of feasible actions. At other times, laypeople may feel that they can exercise the political clout needed to make some options happen, whereas the experts feel constrained to doing what they are paid for. In still other cases, both groups may be silent about very large problems because they see no options.

LAYPEOPLE AND EXPERTS SEE
THE FACTS DIFFERENTLY

There are, of course, situations in which disputes between laypeople and experts cannot be traced to disagreements about objectivity, terminology, problem definitions, process, or feasibility. Having eliminated those possibilities, one may assume the two groups really do see the facts of the matter differently. Here, it may be useful to distinguish between two types of situations: those in which laypeople have no source of information other than the experts, and those in which they do. The reasonableness of disagreements and the attendant policy implications look quite different in each case.

How might laypeople have no source of information other than the experts, and yet come to see the facts differently? One way is for the experts' messages not to get through intact, perhaps because: (1)

The experts are unconcerned about disseminating their knowledge or hesitant to do so because of its tentative nature; (2) only a biased portion of the experts' information gets out, particularly when the selection has been influenced by those interested in creating a particular impression; (3) the message gets garbled in transmission, perhaps due to ill-informed or sensationalist journalists; or (4) the message gets garbled upon reception, either because it was poorly explicated or because recipients lacked the technical knowledge needed to understand the message (Friedman, 1981; Hanley, 1980; Nelkin, 1977). For example, Lord Rothschild (1978) has noted that the BBC does not like to trouble its listeners with the confidence intervals surrounding technical estimates.

A second way of going astray is to misinterpret not the substance, but the process of the science. For example, unless an observer has reason to believe otherwise, it might seem sensible to assume that the amount of scientific attention paid to a risk is a good measure of its importance. Science can, however, be more complicated than that, with researchers going where the contracts, limelight, blue-ribbon panels, or juicy controversies are. In that light (and in hindsight), science may have done a disservice to public understanding by the excessive attention it paid to saccharin ("scientists wouldn't be so involved if this were not a major threat").

A second aspect of the scientific process that may cause confusion is its frequent disputatiousness. It may be all too easy for observers to feel that "if the experts can't agree, my guess may be as good as theirs" (Handler, 1980). Or, they may feel justified in picking the expert of their choice, perhaps on spurious grounds, such as assertiveness, eloquence, or political views. Indeed, it may seldom be the case that the distribution of lay opinions on an issue does not overlap some of the distribution of expert opinions. At the other extreme, laypeople may be baffled by the veil of qualifications that scientists often cast over their work. All too often, audiences may be swayed more by two-fisted debaters (eager to make definitive statements) than by two-handed scientists (saying "on the one hand X, on the other hand Y," in an effort to achieve balance).

In each of these cases, the misunderstanding is excusable, in the sense that it need not reflect poorly on the public's intelligence or on its ability to govern itself. It would, however, seem hard to justify using the public's view of the facts instead of or in addition to the experts' view. A more reasonable strategy would seem to be attempts at education. These attempts would be distinguished from attempts at propaganda by allowing for two-way communication, that is, by being open to the possibility that even when laypeople appear misinformed, they may still have defensible reasons for seeing things differently than do the experts.

For laypeople to disagree reasonably, they would have to have some independent source of knowledge. What might that be? One possibility is that they have a better overview on scientific debates than do the active participants. Laypeople may see the full range of expert opinions and hesitations, immune to the temptations or pressures that actual debaters might feel to fall into one camp and to discredit skeptics' opinions. In addition, laypeople may not feel bound by the generally accepted assumptions about the nature of the world and the validity of methodologies that every discipline adopts in order to go about its business. They may have been around long enough to note that many of the confident scientific beliefs of yesterday are confidently rejected today (Frankel, 1974). Such lay skepticism would suggest expanding the confidence intervals around the experts' best guess at the size of the risks.

Finally, there are situations in which the public, as a result of its life experiences, is privy to information that has escaped the experts (Brokensha et al., 1980). To take three examples: (1) The MacKenzie Valley Pipeline (or Berger) Inquiry discovered that natives of the far North knew things about the risks created by ice-pack movement and sea-bed scouring that were unknown to the pipeline's planners (Gamble, 1978); (2) postaccident analyses often reveal that the operators of machines were aware of problems that the designers of those machines had missed (Sheridan, 1980); and (3) scientists may shy away from studying behavioral or psychological effects (e.g., dizziness, tension) that are hard to measure, and yet still are quite apparent to the individuals who suffer from them. In such cases, lay perceptions of risk should influence the experts' risk estimates (Cotgrove, 1982; Wynne, 1983).

SUMMARY

It is tempting to view others in simplistic terms. Cognitively, one can save mental effort by relying on uncomplicated labels like "the hysterical public" or "the callous experts." Motivationally, properly chosen labels can affirm one's own legitimacy. By the same token, such interpretations can both obstruct the understanding of conflicts (by blurring significant distinctions) and hamper their resolution (by bolstering self-serving characterizations). The following section begins by explaining the consequences of such stereotyping for risk communication by discussing the sort of communication strategies that can follow from simplistic interpretations of the controversy. It continues to outline principles for more complex strategies. These can inform both those designing communications programs and those receiving them.

V
STRATEGIES FOR RISK COMMUNICATION

CONCEPTS OF RISK COMMUNICATION

Risk communication is a collective noun for a variety of procedures expressing quite different attitudes toward the relationship between a society's laypeople and its technical-managerial elite (Covello et al., 1986). At one extreme lies the image of an inactive public docilely waiting for the transmission of vital information from those who know better. Within this perspective, the communication process involves a source, a channel, and a receiver (to use one set of technical terms common among social scientists). Although conceptually simple, this characterization still forces one to consider myriad details about each component. For example (Hovland et al., 1953): How well trusted is the source? Is it a corporate entity, capable of speaking with a single voice, or does it sometimes contradict itself? How much experience and language does the source share with the receivers? How much time does it have to prepare its messages? What are the legal restrictions on how much it can say?

At the other extreme lie highly interactive images of the communication process, in which the public shares responsibility for the social management of risks. Such processes, which require exchanges of information, could, in principle, be viewed as special cases of the source–channel–receiver model. However, using that model (and the research associated with it) requires bearing in mind the notion that these "receivers" are actively shaping the messages that they receive and perhaps even the research conducted in order to create the substance of those messages (Kasperson, 1986).

One way of diagnosing the nature of specific risk communication processes is in terms of the philosophies that guide those who design them. The following discussion describes some generic strategies in terms of their strengths and limitations. The discussion after that considers some more integrative design principles. Together, they are intended to create a framework for responsibly using the more technical material on communication design presented in the final section. That material assumes an understanding of the role of information in the risk management (including communication) process (Johnson and Covello, 1987; Rayner and Cantor, 1987).

SOME SIMPLE STRATEGIES

The technical and policy issues involved in making risk management decisions are complex enough in themselves. Dealing with public perceptions of risks creates an additional level of complexity for risk managers. One possible response to this complexity is to look for some "quick fix" that will deal with the public's needs. Unfortunately for the risk manager, these strategies are both hard to execute well by themselves and unlikely to be sufficient even if they are well executed. At times, these simple solutions seem to reflect a deep misunderstanding of the public's role in risk management, reflecting perhaps a belief that the human element in risk management can be engineered in the same way as mechanical and electronic elements. Undertaken in isolation and with these unrealistic expectations, such strategies can produce mutually frustrating communication programs. The following are some of the more common of these simple strategies for dealing with risk controversies, presented in caricature form to highlight their underlying motivations and inherent limitations.

Give the Public the Facts

The assumption underlying this strategy is that if laypeople only knew as much as the experts, they would respond to hazards in the same way. Undertaken insensitively, this strategy can result in an incomprehensible deluge of technical details, telling the public more than it needs to know about specific risk research results, and much less than it needs to know about the quality of the research (and about how to make the decisions that weigh most heavily on its mind). Concentrating communications on the transmission of information also ignores the possibility that there are legitimate differences between the public and the experts regarding either the goals or the facts of risk management.

Sell the Public the Facts

The premise here is that the public needs persuasion, rather than education. It often follows the failure of an information campaign to win public acceptance for a technology. Undertaken heavy-handedly, this approach may amount to little more than repeating more loudly (or fancily) messages that the public has already rejected. Here, as elsewhere, obvious attempts at manipulation can breed resentment.

Give the Public More of What It Has Gotten in the Past

The underlying assumption here is that the public will accept in the future the kinds of risks that it has accepted in the past. If true, then what the public wants (and will accept) can be determined simply by examining statistics showing the risk–benefit trade-offs involved in existing technologies. This "revealed preference" philosophy ignores the fact, consistently revealed by opinion polls showing great public support for environmental regulations, that people are unhappy with how risks have been managed in the past. The risks that people have tolerated are not necessarily acceptable to them. As a result, giving them more of the same means enshrining past inequities in future decisions. In principle, this approach attaches no importance to educating the public, to creating a constituency for risk policies, or to involving the public in the political process. It seems to respect the public's wishes, while keeping the public itself at arm's length.

Give the Public Clear-Cut, Noncontroversial Statements of Regulatory Philosophy

The assumption underlying this family of approaches is that people do not want facts, but instead the assurance that they are being protected. That is, whatever the risks may be, they are in line with government policy. Examples in the United States include the Delaney clause, prohibiting carcinogenic additives in foods, and the Nuclear Regulatory Commission's "safety goals for nuclear power," describing how risky it will allow the technology to be. Each policy is stated in terms of levels of acceptable risk, as though laypeople are too unsophisticated to understand, in the context of technology management, the sort of risk–benefit trade-offs that they routinely make in everyday life, such as when they undergo medical treatments or pursue hazardous occupations. Moreover, such simple statements provide little guidance for many real situations—by denying the complexity of the (risk–benefit) decisions that needed to be made. If perceived as hollow, then they will do little to reassure the public.

Let the Marketplace Decide

Another hope for risk communication is that risks will be understood when communicated in the context of specific consumer decisions. One variant on this approach is the claim that reducing government regulation will allow people to decide independently what risks they are willing to accept, with the courts addressing any excesses. A second variant is providing quantitative risk information along with goods and drugs. It makes optimistic assumptions regarding laypeople's ability to know enough to fend for themselves with all life's risks. The assumption of personal responsibility and the motivation to get it right are meant to prompt efficient acquisition and understanding. It assumes that people will recognize the limits to their risk perceptions and grasp the risk information presented to them. A threat to any approach emphasizing self-reliance is that people might not want to defend their own welfare when it comes to health and safety, especially where risks have long latencies and it is impossible to prove the source of a health risk (and obtain redress).

Put Risk Managers on the Firing Line

The assumption underlying this strategy is that what the public needs in order to understand risk issues is a coherent story from a single credible source. Examples might include the Nuclear Regulatory Commission's reliance on a single spokesperson as the Three Mile Island incident wore on and the assumption of center stage by the president of Union Carbide after the chemical gas leak in Bhopal, India. This strategy can reduce the confusion created by incomplete conflicting messages, although only if the manager has good communication skills or is sensitive to listeners' information needs; that is, there must be both substance and style. Oversimplifications, misrepresentations, and unacceptable policies are just that, even if they come from a nice guy. This approach can also create a bottleneck for understanding the public's concerns to the extent that the single source of information must also be the single recipient.

Involve Local Communities in Resolving Their Own Risk Management Problems

This approach assumes that people will be flexible and realistic about trade-offs when they see—and have responsibility for—the big picture. Such an approach can founder when the community lacks real decision-making authority or the technical ability to understand its alternatives. It may also founder when those alternatives accept perceived past inequities (e.g., reduce chronic poverty by accepting a hazardous waste dump) or are of the jobs-versus-health variety that people expect government to help them resolve. Ensuring the informed consent of the governed for the risks to which they are exposed is a laudable goal. However, its achievement requires that people have tolerable choices, adequate information, and the ability to identify which course of action is in their own best interests.

CONCEPTUALIZING COMMUNICATION PROGRAMS

Despite their flaws, these simple strategies all have some merit. It is important to give people the facts and to be persuasive when the facts do not speak for themselves or when existing prejudices must be overcome. It is also important to maintain some consistency with past risk management decisions, to expound clear policies, to exploit the wisdom of the marketplace, to encourage direct communication between risk managers and the public, and to give communities meaningful control over their own destinies. The problem is that each strategy oversimplifies the nature of risk issues and the public's involvement with them. When risk managers pin unrealistic hopes on such strategies, then the opportunity to address the public's needs more comprehensively is lost. When these hopes are not met, the frustration that follows is often directed at the public.

It is both unfair and corrosive for the social fabric to criticize laypeople for responding inappropriately to risk situations for which they were not adequately prepared. It is tragic and dangerous when members of our technical elite feel that they have devoted their lives to creating a useful technology (e.g., nuclear power) only to have it rejected by a foolish and unsophisticated public. Likewise, it is painful and unfortunate when the public labels those elites as evil and arrogant.

Risk management requires allocating resources and making trade-offs between costs and benefits. Thus, it inherently involves conflicts. Both the substance and the legitimacy of these conflicts are obscured, however, when the participants come to view them as struggles between the forces of good and evil, or of wisdom and stupidity. Effective solutions will have to be respectful solutions, recognizing both the legitimacy and complexity of the public's perspective, giving it no more and no less credit for reasonableness than it deserves.

How can the preceding observations about risk perceptions (and the research literature from which they were drawn) be used to design better procedures for dealing with risk controversies?

One necessary starting point is a detailed consideration of the nature of the risk that the public must understand. That consideration must cover not only the best available technical estimates for the magnitude of the risk, but also the best available psychological evidence on how people respond to that kind of risk. Research has shown, for example, that people have special demands for safety— and reassurance—when risks are perceived to have delayed effects or catastrophic potential, and when risks appear to be poorly understood or out of people's personal control (Slovic, 1986; Vlek and Stallen, 1980, 1981; von Winterfeldt et al., 1981). Such risks are likely to grab people's attention and create unrest until they can be put in some acceptable perspective. They demand greater communication resources, with particular attention devoted to creating an atmosphere of trust. Perhaps paradoxically, people may need to be treated with the greatest respect in those situations in which they may seem most emotional (or most human) (Eiser, 1982; Weinstein, 1987).

A second necessary starting point is a detailed description of how information about risk can reach people (Johnson and Covello, 1987; Rubin and Sachs, 1973; Schudson, 1978). Such information may be the result of accidents at various distances away and attributed to various causes (e.g., malfunctions, human error, sabotage) or of mere "incidents," such as newspaper exposés, siting controversies, false alarms, or government inquiries. Proactively, this analysis will show the opportunities for reaching people. For example, is there a chance to educate at least some of the public in advance, or can one only prepare materials for times of crisis? Reactively, this analysis should help one anticipate what people will already know (or believe) when the time comes for systematic communication. It may show that people are buffeted by confusing, contradictory, and erroneous messages—or that they have some basic understanding within which they can integrate new information. In any case, communication must build on people's current mental representation of the technology—even if its first step is to challenge inappropriate beliefs and enhance people's ability to examine future information more critically.

Knowing what people do know allows a systematic analysis of what they need to know—the next point of departure in communicating with the public. In some cases, crude estimates of a technology's risks and benefits may be enough; in other cases, it may be important

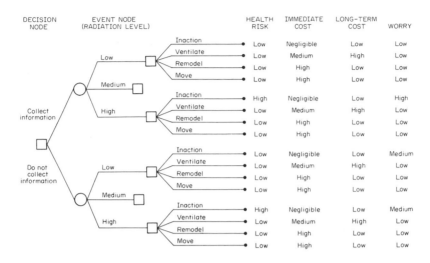

DECISION NODE	EVENT NODE (RADIATION LEVEL)			HEALTH RISK	IMMEDIATE COST	LONG-TERM COST	WORRY
			Inaction	Low	Negligible	Low	Low
		Low	Ventilate	Low	Medium	High	Low
			Remodel	Low	High	Low	Low
			Move	Low	High	Low	Low
	Medium						
Collect information	High		Inaction	High	Negligible	Low	High
			Ventilate	Low	Medium	High	Low
			Remodel	Low	High	Low	Low
			Move	Low	High	Low	Low
			Inaction	Low	Negligible	Low	Medium
Do not collect information	Low		Ventilate	Low	Medium	High	Low
			Remodel	Low	High	Low	Low
			Move	Low	High	Low	Low
	Medium						
	High		Inaction	High	Negligible	Low	Medium
			Ventilate	Low	Medium	High	Low
			Remodel	Low	High	Low	Low
			Move	Low	High	Low	Low

FIGURE V.1 The radiation hazard in homes from the residents' perspective. SOURCE: Svenson and Fischhoff, 1985.

to know how a technology operates. The needs depend on the problems that the public is trying to solve: what to do in an emergency; how to react in a siting controversy; whether to eat vegetables, or whether to let their children do so; and so on. Perhaps the most efficient description would be in the terms of decision theory, such as the simple decision tree in Figure V.1, depicting the situation faced by the head of a household deciding whether to test for domestic radon accumulations. Such descriptions allow one to determine how sensitive these decisions are to different kinds of information, so that communication can focus on the things that people really need to know.

Producing comparable descriptions for the different actors in a risk management episode will help clarify sources of disagreement among them. Often the risk managers' decision problem (e.g., whether to ban EDB) will be quite different from the public's decision problem (e.g., whether to use blueberry muffin mix). For example, Figure V.2 shows the key decision problem that might face risk managers concerned about radon: what standard to set as expressing a tolerable level of exposure. The critical outcomes of this decision are quite different from those associated with the residents' focal decision of whether to test their homes for radon (Figure V.1). Failure

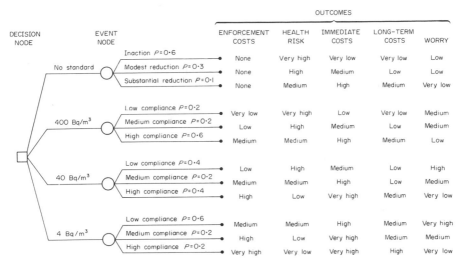

FIGURE V.2 The radiation hazard in homes from the authorities' perspective.
SOURCE: Svenson and Fischhoff, 1985.

to address the public's information needs is likely to leave them frustrated and hostile. Failure to address the managers' own problems is likely to leave their eventual actions inscrutable. For telling their own story, the managers need a protocol that will ensure that all of the relevant parts get out, including what options they are legally allowed to consider, how they see the facts, and what they consider to be the public interest. Such comprehensive accounts are often absent from the managers' public pronouncements, preventing the public from responding responsibly and suggesting that the managers failed to consider the issues fully. The procedures offered in Section II as ways for the public (or the media) to discover what risk issues are all about might also be used proactively as ways to tell the public (or the media) directly about those risks.

After determining what needs to be said, risk managers can start worrying about how to say it. A common worry is that the public will not be able to understand the technical details of how a technology operates. Where those details are really pertinent, the services of good science writers and educators may be needed. Perhaps a more common problem is making the basic concepts of risk management clear. Just what is a one-in-a-million chance? What does it mean to protect wastes for a hundred generations? Must we inevitably set a value on human life when resources are allocated for risk reduction?

The psychological research described above has shown the difficulty of these concepts; it is beginning to show ways to communicate them meaningfully. The research base for addressing these obstacles to understanding is described in the next section.

Adopting such a deliberative approach to characterizing people's needs would help avoid the inadvertent insensitivity found in the Institute of Medicine's (1986) report, *Confronting AIDS*. The report noted, somewhat despairingly, that only 41 percent of the general public knew that AIDS was caused by a virus. Yet, although this fact is elemental knowledge for medical researchers, it has relatively little practical importance for laypeople—in the sense that one would be hard pressed to think of any real decision whose resolution hinged on knowing that AIDS was a virus. Laypeople interested in a deep understanding of the AIDS problem ought to know this fact. However, it is irrelevant to laypeople satisfied just to make reasonable decisions regarding AIDS. Such insensitivity is socially damaging insofar as it demeans the public in the eyes of the experts and prompts the provision of seemingly irrelevant communications.

Another example of this insensitivity to the needs of message recipients can be found in the advice literature about sexual assault (Morgan, 1986). Much of the research is performed and communicated without consideration for women's decision-making needs (Furby and Fischhoff, in press). Most studies concentrate on significance levels, whereas what women need is reliable information on effect size. That is, women need to know not only whether a strategy makes a difference, but how much of a difference. A second form of insensitivity to women's decision-making needs is that few studies collect data on the temporal order of strategies and consequences. As a result, although if greater physical resistance by women were associated with greater violence by men, one would not know which causes which. A third form of insensitivity can be found in recommendations telling women how to respond to different kinds of assailants, without considering whether women can even make such diagnoses under real-life conditions or without reporting the overall prevalence (or "base rates") of the different assailant types, an essential piece of information for making any diagnosis. Finally, some studies actually made the "base-rate fallacy" (Bar-Hillel, 1980; Kahneman and Tversky, 1972), concluding, say, that screaming is more effective than fighting because, among women who escape, 80 percent do the former and only 20 percent do the latter.

Taking the details of risk perceptions seriously means reconciling ourselves to a messy process. In managing risks, society as a whole is slowly and painfully learning how to make deliberative decisions about very difficult issues. Avoiding frustration with the failures and with the public that seems responsible for them will help us keep the mental health and mutual respect needed to get through it all.

EVALUATING COMMUNICATION PROGRAMS

Testing Risky Treatments

If they were creating risks rather than explaining them, risk communicators would be subject to various political, legal, and social constraints. If the treatment involved a medical intervention, then there would be a comparable tangle of restrictions. What analogous responsibilities are incumbent on those who treat others with information?

A minimal requirement might be that a communication have positive expected value. That is, its anticipated net effect should be for the good, considering the magnitude and likelihood of possible consequences. Releasing a communication program that flunked this test would be like authorizing a drug with uncompensated side effects.

A minimal standard of proof for passing this minimal test is expert judgment. Thus, a communication technique could be approved if it were "generally regarded as safe" and seemed likely to be at least somewhat effective. Such reliance on experts' intuitions creates the same discomfort as comparable proposals for grandparenting existing drugs or additives because they are familiar and appear to be safe. How do we know they work? Might negative effects simply have escaped notice or measure? Just what do these experts know? Can they be trusted?

More convincing would be empirical evidence from a basic science of risk communication providing some a priori basis for predicting the effects of particular communications. That evidence could be positive, showing that a communication draws on a demonstrated cognitive ability [e.g., people can understand quantitative probabilities, as long as they are not too small (Beyth-Marom, 1982)]. Or, it could be negative, showing that a communication demands a kind of understanding that is not widely distributed [e.g., people have trouble realizing how the probability of failure accumulates from repeated events, such as using a contraceptive device or being exposed to a disease (Bar-Hillel, 1973)].

More convincing still is evidence from a test of the communication itself, performed with individuals like its ultimate recipients and in a setting like that in which it will ultimately be administered. If that setting must be simulated, then the simulation should capture both those features of the actual communication context that interfere with understanding (e.g., talking to friends during the transmission) and those features that can enhance comprehension (e.g., discussing the transmission with friends) (Turner and Martin, 1985).

Evaluative Criteria

Performing an evaluation requires a clear, operable definition of the consequences to be desired and avoided. With medical treatments, identifying the consequences is usually a straightforward process—they are various possible health effects, some good and some bad. What might be more complicated is measuring some of the effects (e.g., those involving delayed consequences) and determining their relative importance. Although medical personnel and their clients are likely to agree about which outcomes are good and which are bad, they need not agree about how good and how bad the outcomes are. For example, they might feel differently about trade-offs between short- and long-term effects or between changes in quality of life and in expected longevity (McNeil et al., 1978). As a result, even after a definitive evaluation, there may be no universal recommendation. A well-understood treatment might be right for some people, but wrong for others.

In evaluating communication programs, similar issues arise, although with a few additional wrinkles. Potential consequences must still be identified. However, the set seems less clearly defined. There are the good and bad health effects, but they may be hard to observe. If a communication causes undue concern, then there may be stress-related effects, but they tend to be quite diffuse (e.g., a few more cases of child abuse, depression, divorce, and so on, scattered through the treated population) (Elliot and Eisdorfer, 1982). On the other side of the ledger, if people do engage in health-enhancing behavior, then the influence of the focal communication must be isolated from that of other information sources (including, perhaps, continued rumination about an issue).

Difficulties in observing the effects of ultimate interest may divert attention to more observable effects closer to the treatment.

One possibility that arises with communication programs (unlike conventional medical treatments) is assessing comprehension of the message. If people have not understood the message, then an appropriate response seems unlikely. The simplest test of comprehension might be remembering the facts of a message. Those recipients who pass it would, however, still have to be tested for whether they are able to use those remembered facts in their decision making. Those who fail the test would still have to be tested for whether they have heard the message, but chose to reject it. Rejection might mean distrusting the source's competence or its motives. That is, the communicators may not seem to know what they are talking about or they may seem inadequately concerned about the recipients' welfare.

Setting Objectives for Communication Programs

It is accepted wisdom that program planning of any sort ought to begin with an explicit statement of objectives, in the light of which a program's elements can be selected and its effects evaluated. Figure V.3 offers one conceptualization of risk communication programs, categorized according to their primary objective.

According to Covello et al. (1986:172–173):

> In the real world, these four types of risk communication tasks overlap substantially, but they still can be conceptually differentiated. The task of informing and educating the public can be considered primarily a non-directive, although purposeful, activity aimed at providing the lay public with useful and enlightening information. In contrast, both the task of encouraging behavior change and personal protective action and that of providing disaster warnings and emergency information can be considered primarily directive activities aimed at motivating people to take specific types of action. These three tasks, in turn, differ from the task of involving individuals and groups in joint problem solving and conflict resolution, in which officials and citizens exchange information and work together to solve health and environmental problems.

As can be seen from Figure V.3, much risk communication is initiated with the communicators' benefit foremost in mind. For example, the sponsors of a technology may wish to reassure a recalcitrant and alarmed public about its safety. If the public's worry is really unwarranted, then everyone comes out ahead: The technology will get a fairer shake and the public will be relieved of an unnecessary worry. The crucial question is what constitutes "unwarranted" concern. One possible definition is exaggerating the magnitude of the risk (or underestimating the magnitude of accompanying benefits).

TYPE 1: **Information and Education**

    o Informing and educating people about risks and risk assessment in general.

        EXAMPLE: statistical comparisons of the risks of different energy production technologies.

TYPE 2: **Behavior Change and Protective Action**

    o Encouraging personal risk-reduction behavior.

        EXAMPLE: advertisements encouraging people to wear seat belts.

TYPE 3: **Disaster Warnings and Emergency Information**

    o Providing direction and behavioral guidance in disasters and emergencies.

        EXAMPLE: sirens indicating the accidental release of toxic gas from a chemical plant.

TYPE 4: **Joint Problem Solving and Conflict Resolution**

    o Involving the public in risk management decision-making and in resolving health, safety, and environmental controversies.

        EXAMPLE: public meetings about a possible hazardous waste site.

FIGURE V.3   A typology of risk communication objectives. SOURCE: Covello et al., 1986.

In such cases, straight information messages might help. However, they need to be designed with an eye to implicit as well as explicit content. For example, if they are perceived as insistently repeating that "the risk is only X" (or that "the benefit is really Y"), then recipients may read between the lines, "and that ought to be good enough for you." Communicators may convince themselves about the rectitude of such implicit messages, feeling that expert knowledge about the size of risks generalizes to expert knowledge about their acceptability.

Certainly, people should be better off with better information. However, even well-informed people may dislike a technology if they feel that its benefits (to them) are not commensurate with its risks (to them), or that those benefits are substantially lower than the benefits enjoyed by a technology's sponsor. Honest communications should help people reach such determinations. As a result, neither the senders nor the recipients of messages should be faulted if more information leads to more opposition.

An alternative definition of "unwarranted concern" is "larger than the concern associated with hazards having equivalent risk." In more sophisticated versions, the comparison might be with concern over hazards having an equivalent relationship between risks and benefits. A popular contribution to the risk literature a decade ago was lists of disparate risks, chosen so that most were, arguably, accepted by most people (Cohen and Lee, 1979; Crouch and Wilson, 1982). The lists would also contain some favored technology (e.g., nuclear power) that should seemingly be accepted, by whatever criterion led to the acceptance of the other risks in the list. Such lists might, if thoughtfully assembled, help to educate readers' intuitions about the relative magnitude of different risks and the nature of very small risks (e.g., 10^{-6}), such as often appear in such lists. However, even recipients who accept the general idea of consistency that underlies such claims need not accept the particular form of consistency implied by the list (Covello et al., 1988). They may not endorse the particular definition of risk used in the list; they may not feel that all currently accepted (or tolerated or endured) risks are actually acceptable (in the sense that they have agreed voluntarily to the hazards bearing those risks and would not want lesser risks if those were available at a reasonable price). Nor need people accept even the weaker consistency claim that they should not worry more about any hazard than they worry about hazards that they believe to have greater risks. Section III discusses some of people's reasons for ignoring admittedly large hazards.

Comprehension of risk messages is seldom the consequence that is ultimately of interest. Rather, it is a potentially observable surrogate for actual improvements in well-being. A step closer to that consequence would be evidence that recipients of a message had connected their perception of its contents with the course(s) of action in their own best interests (i.e., what a decision theorist would prescribe, given recipients' definition of the situation). For achieving this goal, recipients could be left to their own devices, or they might be provided some help in connecting their beliefs and values with possible actions.

Assuming that it can be done in a neutral (noncoercive) way, providing such help changes the nature of the relationship. Rather than one party administering an informational treatment to another, the treater becomes more of an aide and servant. One particular expression of the change emerges in situations in which a communicator wishes to claim that people have given "informed consent" to the risks described in a communication (P.S. Appelbaum et al., 1987). That claim should interest people exposed to the risks only if it changes their bargaining position vis-à-vis the creator of the risks (e.g., "what's it worth to you for me to sign this release?" or "does that mean that I can force you to give me more information about potential adverse health effects?"). What people should care about is identifying the best choice of action. A communication serves that end if it provides people with the information that they need in a form that they can use. In this light, informed consent may be claimed when people have chosen the best possible course of action for themselves.

These criteria for evaluating risk communication, like those typically invoked for evaluating medical treatments, are focused on direct effects of simple interventions. However, any treatment is but one in a series (at least for those who survive). For example, treatment with an antibiotic might cause no immediate adverse side effects, but might still create an allergic condition that reduces the set of possible treatments for future maladies. Good communication can enhance recipients' actual and perceived ability to understand a risky world and deal with it effectively. Poor communication can do the opposite, reducing recipients' confidence in their own competence to manage the risks in their lives. Just as emotional involvement can impair understanding of the content of messages, so can misunderstanding messages produce unproductive emotions.

Institutional Controls

If risk communications were viewed as treatments, then they might also "enjoy" an institutional context like that created for medical treatments. One component might be review panels to scrutinize the protocols for testing or running communication programs. Such panels might both ensure that programs use suitable evaluation criteria (e.g., reflecting both senders' and recipients' needs) and examine messages for attempts to coerce or misinform. Review panels might also provide guidance on ethical issues. For example, if there is a commonly accepted "best" way to convey a certain kind of information, can one legitimately substitute new, experimental methods? How would that decision change as a function of the kind of testing that the accepted method had undergone? Or, what should be done with messages telling people that they are powerless to affect their fate (e.g., they have been exposed to a carcinogen with irreversible effects, such as asbestos)? Recipients' natural concern over the risk could be aggravated by the feeling of helplessness, especially if the risk is perceived as having been imposed by someone else without providing proper consent or compensation. Do senders have a responsibility to provide counseling for those upset by their messages? Might they even restrict dissemination? How would the decision about the communication process change if the information would help recipients (or others) to mobilize their resources in responding to other hazards? If there are only limited resources for communication, who should receive them (e.g., those at greatest risk, those most responsive to available communication techniques, or those most accessible)?

The institutional context for medical treatments attempts not only to ensure that they are delivered properly, but also to address possible failures. Lists of counterindications accompany many treatments. Physicians are always on stand-by, ready to ameliorate the side effects of their treatments. Various mechanisms exist for collecting and disseminating (good and bad) experiences, for both veteran and experimental treatments. When the rate of side effects is unacceptable, either for a treatment or for a treater, government and professional bodies may stop the exposure. In the background of all these efforts to manage risks lurks the threat of legal proceedings to rectify unmanaged problems (e.g., malpractice and product liability suits). People are more likely to behave well when there are strong social norms for doing so and significant penalties for failure. The desire to be fair to all parties prompts a sharpening of standards.

It took many years to evolve these institutions and standards (many centuries, if one reaches back to Hippocrates). Judging by the various contemporary crises (e.g., malpractice, cost containment), they are still far from perfect. However, those imperfections pale before those of treatments with no such infrastructure. In cases in which an institutional context is created anew for a particular cause, it may be hard to get this degree of balance. For example, right-to-know laws have recently been enacted to ensure that workers receive information about occupational hazards. The laws are intended to help workers protect themselves on the job and to help employers protect themselves in court (by strengthening their claim that workers have given informed consent to bearing the risks). The criteria for evaluating these efforts seem to concentrate more on what is said than on what is understood, raising the threat of overloaded and overly technical messages filling the letter but not the intent of the law. The existence of such threats suggests a tenuous state of affairs for even the more developed areas of risk communication.

SUMMARY

Risk information is an important part of many human activities. Yet it is at most but a part. Understanding its role is essential to giving risk communication programs their basic shape, with appropriate objectives and realistic expectations. Such an analysis can help communicators avoid simplistic strategies that leave recipients, at best, unsatisfied and, at worst, offended by the failure to address their perceived needs. In some cases, these will be for better information; in other cases, they will be for better protection. Only after communication programs are recipient centered in this respect can they productively begin to be recipient centered in the sense of the following section, considering laypeople's strengths and weaknesses in understanding risk information.

VI

PSYCHOLOGICAL PRINCIPLES IN COMMUNICATION DESIGN

Whenever they read a brochure, talk to their neighbors, or observe ominous activities at a local plant in order to understand the risks of a technology, people must rely on the same basic cognitive processes that they use to understand other events in their lives. As mentioned in Section II, the study of such processes is an involved pursuit, with many methodological nuances (like most sciences). To provide some access to the substantive results of such research, here are a number of relatively simple and generally supported statements about behavior. The difficulty in applying them to the prediction of real-life behavior is that life's situations are complex, meaning that various simple behaviors interact in ways that require a subtle analysis to understand.

PEOPLE SIMPLIFY

Most substantive decisions require people to deal with more nuances and details then they can readily handle at any one time. People have to juggle a multitude of facts and values when deciding, for example, whether to change jobs, trust merchants, or protest a toxic landfill. To cope with this information overload, people simplify. Rather than attempting to think their way through to comprehensive, analytical solutions to decision-making problems, people try to rely on habit, tradition, the advice of neighbors (or the media), and on general rules of thumb (e.g., nothing ventured, nothing gained). Rather than consider the extent to which human behavior varies from situation to situation, people describe other people in terms of all-encompassing personality traits, such as being honest, happy, or risk seeking (Nisbett and Ross, 1980). Rather than think precisely about the probabilities of future events, people rely on vague quantifiers, such as "likely" or "not worth worrying about"—terms that are also used differently by different people and by the same individual in different contexts (Beyth-Marom, 1982).

The same desire for simplicity can be observed when people press risk managers to categorize technologies, foods, or drugs as "safe" or "unsafe," rather than treating safety as a continuous variable. It can be seen when people demand convincing proof from scientists who can provide only tentative findings. It can be seen when people attempt to divide the participants in risk disputes into good guys and bad guys, rather than viewing them as people who, like themselves, have complex and interacting motives. Although such simplifications help people cope with life's complexities, they can also obscure the fact that most risk decisions involve gambling with people's health, safety, and economic well-being in arenas with diverse actors and shifting alliances.

ONCE PEOPLE'S MINDS ARE MADE UP, IT IS DIFFICULT TO CHANGE THEM

People are extraordinarily adept at maintaining faith in their current beliefs unless confronted with concentrated and overwhelming evidence to the contrary. Although it is tempting to attribute this steadfastness to pure stubbornness, psychological research suggests that some more complex and benign processes are at work (Nisbett and Ross, 1980).

One psychological process that helps people maintain their current beliefs is feeling little need to look actively for contrary evidence. Why look, if one does not expect that evidence to be very substantial or persuasive? For example, how many environmentalists read *Forbes* and how many industrialists read the Sierra Club's *Bulletin* in order to learn something about risks (as opposed to reading these publications to anticipate the tactics of an opposing side)? A second contributing thought process is the tendency to exploit the uncertainty surrounding apparently contradictory information in order to interpret it as being consistent with existing beliefs. In risk debates, a stylized expression of this proficiency is finding just enough problems with contrary evidence to reject it as inconclusive.

A third thought process that contributes to maintaining current beliefs can be found in people's reluctance to recognize when information is ambiguous. For example, the incident at Three Mile Island would have strengthened the resolve of any antinuclear activist who asked only, "how likely is such an accident, given a fundamentally unsafe technology?", just as it would have strengthened the resolve of any pronuclear activist who asked only, "how likely is the containment of such an incident, given a fundamentally safe technology?" Although a very significant event, Three Mile Island may not have revealed very much about the riskiness of nuclear technology as a whole. Nonetheless, it helped the opposing sides polarize their views. Similar polarization has followed the accident at Chernobyl, with opponents pointing to the "consequences of a nuclear accident" (which come with any commitment to nuclear power) and proponents pointing to the unique features of that particular accident (which are unlikely to be repeated elsewhere, especially considering the precautions instituted in its wake) (Krohn and Weingart, 1987).

PEOPLE REMEMBER WHAT THEY SEE

Fortunately, given their need to simplify, people are quite good at observing those events that come to their attention (and that they are motivated to understand) (Hasher and Zacks, 1984; Peterson and Beach, 1967). As a result, if the appropriate facts reach people in a responsible and comprehensible form before their minds are made up, there is a decent chance that their first impression will be the correct one. For example, most people's primary sources of information about risks are what they see in the news media and observe in their everyday lives. Consequently, people's estimates of the principal causes of death are strongly related to the number of people they know who have suffered those misfortunes and the amount of media coverage devoted to them (Lichtenstein et al., 1978).

Unfortunately for their risk perceptions (although fortunately for their well-being), most people have little firsthand knowledge of hazardous technologies. Rather, what laypeople see most directly are the outward manifestations of the risk management process, such as hearings before regulatory bodies or statements made by scientists to the news media. In many cases, these outward signs are not very reassuring. Often, they reveal acrimonious disputes between supposedly reputable experts, accusations that scientific findings have been distorted to suit their sponsors, and confident assertions that are disproven by subsequent research (Dietz and Rycroft, 1987; MacLean, 1987; Rothman and Lichter, 1987).

PEOPLE CANNOT READILY DETECT OMISSIONS IN THE EVIDENCE THEY RECEIVE

Not all problems with information about risk are as readily observable as blatant lies or unreasonable scientific hubris. Often, the information that reaches the public is true, but only part of the truth. Detecting such systematic omissions proves to be quite difficult (Tversky and Kahneman, 1973). For example, most young people know relatively few people suffering from the diseases of old age; nor are they likely to see those maladies cited as the cause of death in newspaper obituaries. As a result, young people tend to underestimate the frequency of these causes of death, while overestimating the frequency of vividly reported causes, such as murder, accidents, and tornadoes (Lichtenstein et al., 1978).

Laypeople are even more vulnerable when they have no way of knowing about information because it has not been disseminated. In principle, for example, patients could always ask their physicians whether they have neglected to mention any side effects of the drugs they prescribe. Likewise, people could always ask merchants whether there are any special precautions for using a new power tool, or ask proponents of a hazardous facility if their risk assessments have considered operator error and sabotage. In practice, however, these questions about omissions are rarely asked. It takes an unusual turn of mind to recognize one's own ignorance and insist that it be addressed.

As a result of this insensitivity to omissions, people's risk perceptions can be manipulated in the short run by selective presentation. Not only will people not know what they have not been told, but they will not even notice how much has been left out (Fischhoff et al., 1978a). What happens in the long run depends on whether the unmentioned risks are revealed by experience or by other sources of information. When deliberate omissions are detected, the responsible party is likely to lose all credibility. Once a shadow of doubt has been cast, it is hard to erase.

PEOPLE MAY DISAGREE MORE ABOUT WHAT RISK IS THAN ABOUT HOW LARGE IT IS

Given this mixture of strengths and weaknesses in the psychological processes that generate people's risk perceptions, there is no simple answer to the question "how much do people know and understand?" The answer depends on the risks and on the opportunities that people have to learn about them.

One obstacle to determining what people know about specific risks is disagreement about the definition of risk. (See Sections II and III for more complete discussions of different possible definitions of risk and other terms.) If laypeople and risk managers use the term risk differently, then they can agree on the facts about a specific technology but still disagree about its degree of riskiness. Several years ago, the idea circulated in the nuclear power industry that the public cared much more about multiple deaths from large accidents than about equivalent numbers of casualties resulting from a series of small accidents. If this assumption were valid, then the industry would be strongly motivated to remove the threat of such large accidents. If removing the threat proved impossible, then the industry could argue that a death is a death and that in formulating social policy it is totals that matter, not whether deaths occur singly or collectively.

There were never any empirical studies to determine whether this was really how the public defined risk. Subsequent studies, though, have suggested that what bothers people about catastrophic accidents is the perception that a technology capable of producing such accidents cannot be very well understood or controlled (Slovic et al., 1984). From an ethical point of view, worrying about the uncertainties surrounding a new and complex technology such as nuclear power is quite a different matter than caring about whether a fixed number of lives are lost in one large accident rather than in many small accidents.

PEOPLE HAVE DIFFICULTY DETECTING INCONSISTENCIES IN RISK DISPUTES

Despite their frequent intensity, risk debates are typically conducted at a distance (Hance et al., 1988; Mazur, 1973). The disputing parties operate within self-contained communities and talk principally to themselves. Opponents are seen primarily through their writing or their posturing at public events. Thus, there is little opportunity for the sort of subtle probing needed to discover basic differences in how the protagonists think about important issues, such as the meaning of key terms or the credibility of expert testimony. As a result, it is easy to misdiagnose one another's beliefs and concerns.

The opportunities for misunderstanding increase when the circumstances of debate restrict candor. For example, some critics of nuclear power actually believe that the technology can be operated with reasonable safety. However, they oppose it because they believe that its costs and benefits are distributed inequitably. Although they might like to discuss these issues, critics find that public hearings about risk and safety often provide them with their only forum for venting their concern. If they oppose the technology, then they are forced to do so on safety grounds, even if this means misrepresenting their perceptions of the actual risk.

Individuals also have difficulty detecting inconsistencies in their own beliefs or realizing how simple reformulations would change their perspective on issues. For example, most people would prefer a gamble with a 25 percent chance of losing $200 (and a 75 percent chance of losing nothing) to a gamble with a sure loss of $50. Most of the same people would also buy a $50 insurance policy to protect against such a loss. What they will do depends on whether the $50 is described as a sure loss or as an insurance premium. As a result, one cannot predict how people will respond to an issue without knowing how they will perceive it, which depends, in turn, on how it will be presented to them by merchandisers, politicians, or the media.

Thus, people's insensitivity to the importance of how risk issues are presented exposes them to manipulation. For example, a risk might seem much worse when described in relative terms than in absolute terms (e.g., doubling their risk versus increasing that risk from 1 in a million to 1 in a half million). Although both representations of the risk might be honest, their impacts would be quite different. Perhaps the only fair approach is to present the risk from both perspectives, letting recipients determine which one (or which hybrid) best represents their world view.

SUMMARY

These statements (and others like them cited elsewhere in this appendix) reduce both complex people and intricate research literatures to necessarily oversimplified summaries. Neither the people nor the literature can be read without their appropriate context. Much of Section II discussed the intricacies of the literature and the sort of conclusions than might be extracted from it. Much of this whole appendix concerns the context for risk perception. Ideally, one would have polished studies of how specific people respond to specific risks, either in messages or in the flesh (or the metal). Those should be the standards for designing and evaluating risk communication programs. In lieu of such studies, such principles are all that we have to go on. They are the stuff of everyday explanations of behavior. They can be enriched, refined, and (sometimes) disqualified by behavioral research.

VII

CONCLUSION

INDIVIDUAL LEARNING

Making decisions about risks is often complex, whether done individually or as part of a larger social-political process. So is dealing with many of life's other decisions, even without obvious risks to health and safety (e.g., choosing a career, a partner, an anniversary present). All these decisions have sets of options to consider, bodies of fact to master, and competing objectives to weigh. Adding to the complexity of these individual decisions is the fact that each of us confronts so many of them—each with its own details and nuances.

Individually and collectively, these decisions present a daunting challenge to identify those courses of action that are in our own best interests. It should not be surprising if people sometimes feel overwhelmed by the panoply of risks thrown at them, sometimes seem to respond suboptimally, and sometimes get angry at those who force them to deal with yet another risk—even if it is associated with a technology bringing considerable benefit.

However, although the substance of these decisions may vary enormously, their common elements mean that there is an opportunity for learning some general lessons from this experience with diverse risks. So, even though few people receive formal training in decision-making methods, life itself can provide an education. People could not make it through life if they had not learned something about the relative riskiness of different activities (e.g., driving at night versus driving during the day, getting polio from vaccine versus getting it while unvaccinated, storing household chemicals under the sink versus storing them out of the reach of children). People would be perennially dissatisfied if they had not acquired some ability to understand and predict their own tastes. A representative democracy could not function if people did not have some ability to evaluate the candor and competence of political candidates and governmental officials. There would not be significant declines in smoking and fat consumption if people were not able to extract personally relevant implications from risk communications.

Some of these accomplishments are documented in the references cited in the preceding sections. Most are also common knowledge (although perhaps not as precisely delineated as they can be in systematic research). Most are also incomplete. Both anecdotal

and systematic observations can point to places where people misestimate risks, mistake their own needs, misjudge public figures, or
misinterpret the message of risk communications. In some cases, this
is because life is not structured for learning. It may not provide
people with prompt, immediate feedback on how well they are doing.
It may discourage them from admitting the need to learn (without
which even the sharpest feedback may have little value).

Under these circumstances, a guide like this can facilitate learning in several ways. One is to provide a structure for thinking about
risk controversies, so as to facilitate identifying common elements
and extracting general lessons. A second is to summarize the lessons
found in the research literature and in the pooled experience of risk
communicators (and communicants). In some cases, these lessons
will confirm readers' expectations; in others, they will suggest alternative interpretations; in still others, they will raise issues that
have not been considered. A third way is to provide annotated references to the research literature that could be consulted for more
detailed treatment of specific risk issues. Making this research generally available in nontechnical terms can help to level the playing
field, by granting equal access to it for all parties to risk controversies
(and not just for those parties with staffs paid to follow the research
literature).

Finally, such a guide can provide some insight into the psychological processes of the parties involved in risk controversies. That
insight can be used directively, by those who must design risk communications and interpret the responses of the public to them. It
can also be used reflectively, by those who wish to clarify the psychological limits to their own participation in risk management. These
groups include nontechnical people concerned about interpreting the
nature of risks, as well as technical people concerned about making
themselves understood to others.

Such understanding has both a "cognitive" and a "motivational"
component (to use psychological jargon for a moment). That is, it
involves both how people think and how people feel. Deciphering
scientific communications can be complicated both by difficulty interpreting strange terms or unfamiliar units (e.g., very small probabilities) and by difficulty coping with one's anger with the risk
communicators (e.g., for their perceived insensitivity or vested interests). Designing such communications can be complicated both
by difficulty interpreting complex social processes and by difficulty

managing one's frustration at being mistrusted and disbelieved. Better risk communication is typically thought of as a largely cognitive enterprise, focused on conveying factual material more comprehensibly. Accomplishing that goal requires an understanding of what aspects of risk conflicts really hinge on scientific facts. If it can be accomplished, then risk conflicts can be focused on areas of legitimate disagreement, without the confusion and frustration generated by the receipt of incomprehensible messages. Such messages both blur the issues and create the feeling that communicators care so little— or live in such a different world—that they cannot communicate in ways that address recipients' needs.

SOCIETAL LEARNING

Sweeping statements about people and society are easy to make, but hard to substantiate. If I were to chance a summary of personal observations from 15 years of working on this topic, it would be that there is increasing sophistication on the part of all concerned. We have better risk science than we had in the past and a better understanding of its limits. We have increasing understanding among risk managers of the need to take public concerns seriously when designing risk policies and among members of the public when deciding which risks to worry about and how to worry about them. We have increasing professionalism in reporting about risk issues and increasing ability to read or view risk stories with a discerning eye.

We also have, however, a long way to go in each of these respects. Moreover, the learning to date has come at a price that creates an obstacle to future progress. People remember their own past mistakes (at least the more obvious ones), which makes them hesitant about future actions. They also remember others' mistakes (at least those from which they think they have suffered), which makes them leery of those others' future actions. It is hard to erase a shadow of doubt or undo the undue impact of first impressions.

As in a social relationship, by the time those involved learn how to get along with a significant other, they may have hurt one another enough that they cannot apply these lessons in that relationship. Unfortunately, industry cannot break off its relationship with its current public (or its current government or current media) and start up with a new, more enlightened one. So, some personal wounds need to heal at the same time as we are collectively addressing new problems.

In addition, old problems continue to aggravate these wounds and to undermine the parties' faith in one another. For example, the question of whether to complete or operate many nuclear reactors is a lingering source of mutual frustration among all involved. The public commitments made by the various parties concerned are such that the conflicts have a life of their own. They may defy reasoned resolution and be almost refractory to the addition of scientific evidence. The strategizing and posturing of the parties may make great sense when viewed as part of a political struggle. Yet when viewed as part of a disciplined debate over risks and benefits, they can strengthen perceptions of a callous industry and hysterical public.

A guide such as this cannot dispel such complex conflicts and emotions. They are natural and legitimate parts of life. It can, however, help to put them in perspective, leaving the conflicts that remain better focused and more productive.

BIBLIOGRAPHY

Alfidi, J. 1971. Informed consent: A study of patient reaction. Journal of the American Medical Association 216:1325–1329.

Appelbaum, P. S., C. W. Lidz, and A. Meisel. 1987. Informed Consent: Legal Theory and Clinical Practice. New York: Oxford University Press.

Appelbaum, R. P. 1977. The future is made, not predicted: Technocratic planners vs. public interests. Society (May/June):49–53.

Applied Management Sciences. 1978. Survey of consumer perceptions of patient package inserts for oral contraceptives. NTIS No. PB-248-740. Washington, D.C.: Applied Management Sciences.

Armstrong, J. S. 1975. Tom Swift and his electric regression analysis machine 1973. Psychological Reports 36:806.

Atkinson, R. C., R. J. Herrnstein, G. Lindzey, and R. D. Luce. 1988. Stevens' Handbook of Experimental Psychology. New York: Wiley Interscience.

Bar-Hillel, M. 1973. On the subjective probability of compound events. Organizational Behavior and Human Performance 9:396–406.

Bar-Hillel, M. 1980. The base rate fallacy in probability judgment. Acta Psychologica 44:211–233.

Barber, W. C. 1979. Controversy plagues setting of environmental standards. Chemical and Engineering News 57(17):34–37.

Barraclough, G. 1972. Mandarins and Nazis. New York Review of Books 19(6):37–42.

Bazelon, D. L. 1979. Risk and responsibility. Science 205(4403):277–280.

Bentkover, J. D., V. T. Covello, and J. Mumpower, eds. 1985. Benefits Assessment: The State of the Art. Dordrecht, Holland: D. Reidel.

Berkson, J., T. B. Magath, and M. Hurn. 1939–1940. The error of estimate of the blood cell count as made with the hemocytometer. American Journal of Physiology 128:309–323.

Beyth-Marom, R. 1982. How probable is probable? Journal of Forecasting 1:257–269.

Bick, T., C. Hohenemser, and R. W. Kates. 1979. Target: Highway risks. Environment 21(2):7–15, 29–38.

Bickerstaffe, J., and D. Peace. 1980. Can there be a consensus on nuclear power? Social Studies of Science 10:309–344.

Bradburn, N. M., and S. Sudman. 1979. Improving Interview Method and Questionnaire Design. San Francisco: Jossey-Bass.

Brokensha, D. W., D. M. Warren, and O. Werner. 1980. Indigenous Knowledge: Systems and Development. Lanham, Md.: University Press of America.

Brookshire, D. S., B. C. Ives, and W. D. Schulze. 1976. The valuation of aesthetic preferences. Journal of Environmental Economics and Management 3:325–346.

Brown, R. 1965. Social Psychology. Glencoe, Ill.: Free Press.

Burton, I., R. W. Kates, and G. F. White. 1978. The Environment as Hazard. New York: Oxford University Press.

Callen, E. 1976. The science court. Science 193:950–951.

Campbell, D. T. 1975. Degrees of freedom and the case study. Comparative Political Studies 8:178–193.

Campbell, D. T., and A. Erlebacher. 1970. How regression artifacts in quasi-experimental evaluations can mistakenly make compensatory education look harmful. In Compensatory Education: A National Debate, Vol. 3, Disadvantaged Child, J. Hellmuth, ed. New York: Brunner/Mazel.

Campen, J. 1985. Benefit-Cost and Beyond. Cambridge, Mass.: Ballinger.

Carterette, E. C., and M. P. Friedman. 1974. Handbook of Perception, Vol. 2. New York: Academic Press.

Chapman, L. J., and J. P. Chapman. 1969. Illusory correlation as an obstacle to the use of valid psychodiagnostic signs. Journal of Abnormal Psychology 74:271–280.

Chemical and Engineering News. 1980. A look at human error. 58(18):82.

Cohen, B., and I. Lee. 1979. A catalog of risks. Health Physics 36:707–722.

Cohen, J. 1962. The statistical power of abnormal-social psychological research: A review. Journal of Abnormal and Social Psychology 65(3):145–153.

Commoner, B. 1979. The Politics of Energy. New York: Knopf.

Conn, W. D., ed. 1983. Energy and Material Resources. Boulder, Colo.: Westview.

Cotgrove, A. 1982. Catastrophe or Cornucopia? The Environment, Politics and the Future. New York: John Wiley & Sons.

Covello, V. T., P. M. Sandman, and P. Slovic. 1988. Risk Communication, Risk Statistics, and Risk Comparisons: A Manual for Plant Managers. Washington, D.C.: Chemical Manufacturers Association.

Covello, V., D. von Winterfeldt, and P. Slovic. 1986. Risk communication: A review of the literature. Risk Abstracts 3(4):171–182.

Crask, M. R., and W. D. Parreault, Jr. 1977. Validation of discriminant analysis in marketing research. Journal of Marketing Research 14:60–68.

Crouch, E. A. C., and R. Wilson. 1982. Risk/Benefit Analysis. Cambridge, Mass.: Ballinger.

Cummings, R. G., D. S. Brookshire, and W. D. Schulze, eds. 1986. Valuing Environmental Goods: An Assessment of the Contingent Valuation Method. Totowa, N.J.: Rowman & Allanheld.

Davidshofer, I. O. 1976. Risk-taking and vocational choice: Reevaluation. Journal of Counseling Psychology 23:151–154.

Davis, J. 1969. Group Performance. Reading, Mass.: Addison-Wesley.

Dietz, T. M., and R. W. Rycroft. 1987. The Risk Professionals. Washington, D.C.: Russell Sage Foundation.

Doern, G. B. 1978. Science and technology in the nuclear regulatory process: The case of Canadian uranium miners. Canadian Public Administration 21:51–82.

Dreman, D. 1979. Contrarian Investment Strategy. New York: Random House.

Driver, B., G. Peterson, and R. Gregory, eds. 1988. Evaluative Amenity Resources. New York: Venture.

Dunlap, T. R. 1978. Science as a guide in regulating technology: The case of DDT in the United States. Social Studies of Science 8:265–285.

Eiser, J. R., ed. 1982. Social Psychology and Behavioral Medicine. New York: John Wiley & Sons.

Elliot, G. R., and C. Eisdorfer. 1982. Stress and Human Health. New York: Springer-Verlag.

Fairley, W. B. 1977. Evaluating the "small" probability of a catastrophic accident from the marine transportation of liquefied natural gas. In Statistics and Public Policy, W. B. Fairley and F. Mosteller, eds. Reading, Mass.: Addison-Wesley.

Feller, W. 1968. An Introduction to Probability Theory and Its Applications, 3d ed., Vol. 1. New York: John Wiley & Sons.

Fineberg, H. V. 1988. Education to prevent AIDS: Prospects and obstacles. Science 239(4840):592–596.

Fischer, D. H. 1970. Historians' Fallacies. New York: Harper & Row.

Fischhoff, B. 1980. For those condemned to study the past: Reflections on historical judgment. In New Directions for Methodology of Behavior Science: Fallible Judgment in Behavioral Research, R. A. Shweder and D. W. Fiske, eds. San Francisco: Jossey-Bass.

Fischhoff, B. 1981. Informed consent for transient nuclear workers. In Equity Issues in Nuclear Waste Management, R. Kasperson and R. W. Kates, eds. Cambridge, Mass.: Oelgeschlager, Gunn and Hain.

Fischhoff, B. 1983. "Acceptable risk": The case of nuclear power. Journal of Policy Analysis and Management 2(4):559–575.

Fischhoff, B. 1984. Setting standards: A systematic approach to managing public health and safety risks. Management Science 30:823–843.

Fischhoff, B. 1985a. Managing risk perceptions. Issues in Science and Technology 2(1):83–96.

Fischhoff, B. 1985b. Protocols for environmental reporting: What to ask the experts. The Journalist (Winter):11–15.

Fischhoff, B. 1985c. Risk analysis demystified. NCAP News (Winter):30–33.

Fischhoff, B. 1987. Treating the public with risk communications: A public health perspective. Science, Technology, and Human Values 12:3–19.

Fischhoff, B. 1988. Judgment and decision making. In The Psychology of Human Thought, R. J. Sternberg and E. E. Smith, eds. New York: Cambridge University Press.

Fischhoff, B., and L. A. Cox, Jr. 1985. Conceptual framework for regulatory benefits assessment. In Benefits Assessment: The State of the Art, J. D. Bentkover, V. T. Covello, and J. Mumpower, eds. Dordrecht, Holland: D. Reidel.

Fischhoff, B., and L. Furby. 1988. Measuring values: A conceptual framework for interpretive transactions with special reference to contingent valuation of visibility. Journal of Risk and Uncertainty 1:147–184.

Fischhoff, B., and D. MacGregor. 1983. Judged lethality: How much people seem to know depends upon how they are asked. Risk Analysis 3:229–236.

Fischhoff, B., and O. Svenson. 1987. Perceived risks of radionuclides: Understanding public understanding. In Radionuclides in the Food Chain, G. Schmidt, ed. New York: Praeger.

Fischhoff, B., L. Furby, and R. Gregory. 1987. Evaluating voluntary risks of injury. Accident Analysis and Prevention 19(1):51–62.

Fischhoff, B. S., Lichtenstein, P. Slovic, S. L. Derby, and R. L. Keeney. 1981. Acceptable Risk. New York: Cambridge University Press.

Fischhoff, B., P. Slovic, and S. Lichtenstein. 1978. Fault trees: Sensitivity of assessed failure probabilities to problem representation. Journal of Experimental Psychology: Human Perception and Performance 4:330–344.

Fischhoff, B., P. Slovic, and S. Lichtenstein. 1980. Knowing what you want: Measuring labile values. In Cognitive Processes in Choice and Decision Behavior, T. Wallsten, ed. Hillsdale, N.J.: Erlbaum.

Fischhoff, B., P. Slovic, and S. Lichtenstein. 1981. Lay foibles and expert fables in judgments about risk. In Progress in Resource Management and Environmental Planning, T. O'Riordan and R. K. Turner, eds. New York: John Wiley & Sons.

Fischhoff, B., P. Slovic, S. Lichtenstein, S. Read, and B. Combs. 1978. How safe is safe enough? A psychometric study of attitudes towards technological risks and benefits. Policy Sciences 9:127–152.

Fischhoff, B., S. R. Watson, and C. Hope. 1984. Defining risk. Policy Sciences 17:123–129.

Fiske, S., and S. Taylor. 1984. Social Cognition. Reading, Mass.: Addison-Wesley.

Frankel, C. 1974. The rights of nature. In When Values Conflict, C. Schelling, J. Voss, and L. Tribe, eds. Cambridge, Mass.: Ballinger.

Friedman, S. M. 1981. Blueprint for breakdown: Three Mile Island and the media before the accident. Journal of Communication 31:116–129.

Furby, L., and B. Fischhoff. In press. Rape self-defense strategies: A review of their effectiveness. Victimology.

Gamble, D. J. 1978. The Berger Inquiry: An impact assessment process. Science 199(4332):946–951.

Gilovich, T., R. Vallone, and A. Tversky. 1985. The hot hand in basketball: On the misperception of random sequences. Cognitive Psychology 17:295–314.

Gotchy, R. L. 1983. Health risks from the nuclear fuel cycle. In Health Risks of Energy Technologies, C. C. Travis and E. L. Etnier, eds. Boulder, Colo.: Westview.

Green, A. E., and A. J. Bourne. 1972. Reliability Technology. New York: Wiley Interscience.

Hackney, J. D., and W. S. Linn. 1984. Human toxicology and risk assessment. In Handbook on Risk Assessment. Washington, D.C.: National Science Foundation.

Hammond, K. R., and L. Adelman. 1976. Science, values and human judgment. Science 194:389–396.

Hance, B. J., C. Chess, and P. M. Sandman. 1988. Improving Dialogue with Communities: A Risk Communication Manual for Government. Trenton: Division of Science and Research Risk Communication Unit, New Jersey Department of Environmental Protection.

Handler, P. 1980. Public doubts about science. Science 208(4448):1093.

Hanley, J. 1980. The silence of scientists. Chemical and Engineering News 58(12):5.

Harris, L. 1980. Risk in a complex society. Public opinion survey conducted for Marsh and McLennan Companies, Inc.

Harriss, R., and C. Hohenemser. 1978. Mercury: Measuring and managing risk. Environment 20(9).

Hasher, L., and R. T. Zacks. 1984. Automatic and effortful processes in memory. Journal of Experimental Psychology: General 108:356–388.

Henrion, M., and B. Fischhoff. 1986. Assessing uncertainty in physical constants. American Journal of Physics 54(9):791–798.

Henshel, R. L. 1975. Effects of disciplinary prestige on predictive accuracy: Distortions from feedback loops. Futures 7:92–196.

Herbert, J. H., L. Swanson, and P. Reddy. 1979. A risky business. Environment 21(6):28–33.

Hershey, J. C., and P. J. H. Schoemaker. 1980. Risk taking and problem context in the domain of losses: An expected utility analysis. Journal of Risk and Insurance 47:111–132.

Hirokawa, R. Y., and M. S. Poole. 1986. Communication and Group Decison Making. Beverly Hills, Calif.: Sage.

Hohenemser, K. H. 1975. The failsafe risk. Environment 17(1):6–10.

Holden, C. 1980. Love Canal residents under stress. Science 208:1242–1244.

Hovland, C. I., I. L. Janis, and H. H. Kelley. 1953. Communication and Persuasion: Psychological Studies of Opinion Change. New Haven, Conn.: Yale University Press.

Hynes, M., and E. Vanmarcke. 1976. Reliability of embankment performance prediction. In Proceedings of the ASCE Engineering Mechanics Division Specialty Conference. Waterloo, Ontario, Canada: University of Waterloo Press.

Ingram, M. J., D. J. Underhill, and T. M. L. Wigley. 1978. Historical climatology. Nature 276:329–334.

Inhaber, H. 1979. Risk with energy from conventional and nonconventional sources. Science 203(4382):718–723.

Institute of Medicine. 1986. Confronting AIDS: Directions for Public Health, Health Care, and Research. Washington, D.C.: National Academy Press.

James, W. 1988. Baseball Abstract. New York: Ballantine.

Janis, I. L., ed. 1982. Counseling on Personal Decisions. New Haven, Conn.: Yale University Press.

Jennergren, L. P., and R. L. Keeney. 1982. Risk assessment. In Handbook of Applied Systems Analysis. Laxenburg, Austria: International Institute of Applied Systems Analysis.

Johnson, B. B., and V. T. Covello, eds. 1987. The Social and Cultural Construction of Risk: Essays on Risk Selection and Perception. Dordrecht, Holland: D. Reidel.

Joksimovich, V. 1984. Models in risk assessment for hazard characterization. In Handbook of Risk Assessment. Washington, D.C.: National Science Foundation.

Joubert, P., and L. Lasagna. 1975. Commentary: Patient package inserts. Clinical Pharmacology and Therapeutics 18(5):507–513.

Kadlec, R. 1984. Field and laboratory event investigation for hazard characterization. In Handbook of Risk Assessment. Washington, D.C.: National Science Foundation.

Kahneman, D., and A. Tversky. 1972. Subjective probability: A judgment of representativeness. Cognitive Psychology 3:430–454.

Kasperson, R. 1986. Six propositions on public participation and their relevance for risk communication. Risk Analysis 6(3):275–281.

Keeney, R. L. 1980. Siting Energy Facilities. New York: Academic Press.

Keeney, R. L., and H. Raiffa. 1976. Decisions with Multiple Objectives: Preferences and Value Tradeoffs. New York: John Wiley & Sons.

Kolata, G. B. 1980. Love Canal: False alarm caused by botched study. Science 208(4449):1239–1242.

Koriat, A., S. Lichtenstein, and B. Fischhoff. 1980. Reasons for confidence. Journal of Experimental Psychology: Human Learning and Memory 6:107–118.

Krohn, W., and P. Weingart. 1987. Commentary: Nuclear power as a social experiment—European political "fall-out" from the Chernobyl meltdown. Science, Technology, and Human Values 12(2):52–58.

Kunce, J. T., D. W. Cook, and D. E. Miller. 1975. Random variables and correlational overkill. Educational and Psychological Measurement 35:529–534.

Kunreuther, H., R. Ginsberg, L. Miller, P. Sagi, P. Slovic, B. Borkan, and N. Katz. 1978. Disaster Insurance Protection. New York: John Wiley & Sons.

Lachman, R., J. T. Lachman, and E. C. Butterfield. 1979. Cognitive Psychology and Information Processing. Hillsdale, N.J.: Erlbaum.

Lakatos, I. 1970. Falsification and scientific research programmes. In Criticism and the Growth of Scientific Knowledge, I. Lakatos and A. Musgrave, eds. New York: Cambridge University Press.

Lanir, Z. 1982. Strategic Surprises. Tel Aviv, Israel: Hakibbutz Hameuchad.

Lave, L. B. 1978. Ambiguity and inconsistency in attitudes toward risk: A simple model. Pp. 108–114 in Proceedings of the Society for General Systems Research Annual Meeting. Louisville, Ky.: Society for General Systems Research.

Lawless, E. W. 1977. Technology and Social Shock. New Brunswick, N.J.: Rutgers University Press.

Lazarsfeld, P. 1949. The American soldier—An expository review. Public Opinion Quarterly 13:377–404.

Levine, M. 1974. Scientific method and the adversary model: Some preliminary thoughts. American Psychologist 29:661–716.

Lichtenstein, S., and B. Fischhoff. 1980. Training for calibration. Organizational Behavior and Human Performance 26:149–171.

Lichtenstein, S., B. Fischhoff, and L. D. Phillips. 1982. Calibration of probabilities: The state of the art. In Judgment Under Uncertainty: Heuristics and Biases, P. Slovic and A. Tversky, eds. New York: Cambridge University Press.

Lichtenstein, S., P. Slovic, B. Fischhoff, M. Layman, and B. Combs. 1978. Judged frequency of lethal events. Journal of Experimental Psychology: Human Learning and Memory 4:551–578.

Lindman, H. G., and W. Edwards. 1961. Supplementary report: Unlearning the gambler's fallacy. Journal of Experimental Psychology 62:630.

Linville, P., B. Fischhoff, and G. Fischer. 1988. Judgments of AIDS Risks. Pittsburgh, Pa.: Carnegie-Mellon University, Department of Social and Decision Sciences.

MacLean, D. 1987. Understanding the nuclear power controversy. In Scientific Controversies: Case Studies in the Resolution and Closure of Disputes in Science and Technology, H. T. Engelhardt, Jr., and A. L. Caplan, eds. New York: Cambridge University Press.

Markovic, M. 1970. Social determinism and freedom. In Mind, Science and History, H. E. Keifer and M. K. Munitz, eds. Albany: State University of New York Press.

Martin, E. 1980. Surveys as Social Indicators: Problems in Monitoring Trends. Chapel Hill: Institute for Research in Social Science, University of North Carolina.

Mazur, A. 1973. Disputes between experts. Minerva 11:243–262.

Mazur, A. 1981. The Dynamics of Technical Controversy. Washington, D.C.: Communications Press.

Mazur, A., A. A. Marino, and R. O. Becker. 1979. Separating factual disputes from value disputes in controversies over technology. Technology in Society 1:229–237.

McGrath, P. E. 1974. Radioactive Waste Management: Potentials and Hazards From a Risk Point of View. Report EUR FNR-1204 (KFK 1992). Karlsruhe, West Germany: US-EURATOM Fast Reactor Program.

McNeil, B. J., R. Weichselbaum, and S. G. Pauker. 1978. The fallacy of the 5-year survival rate in lung cancer. New England Journal of Medicine 299:1397–1401.

Morgan, M. 1986. Conflict and confusion: What rape prevention experts are telling women. Sexual Coercion and Assault 1(5):160–168.

Murphy, A. H., and B. G. Brown. 1983. Forecast terminology: Composition and interpretation of public weather forecasts. Bulletin of the American Meteorological Society 64:13–22.

Murphy, A. H., and R. L. Winkler. 1984. Probability of precipitation forecasts. Journal of the American Statistical Association 79:391–400.

National Research Council. 1976. Surveying Crime. Washington, D.C.: National Academy Press.

National Research Council. 1982. Survey Measure of Subjective Phenomena. Washington, D.C.: National Academy Press.

National Research Council. 1983a. Priority Mechanisms for Toxic Chemicals. Washington, D.C.: National Academy Press.

National Research Council. 1983b. Risk Assessment in the Federal Government: Managing the Process. Washington, D.C.: National Academy Press.

Nelkin, D. 1977. Technological Decisions and Democracy. Beverly Hills, Calif.: Sage.

Nelkin, D., ed. 1984. Controversy: Politics of Technical Decisions. Beverly Hills, Calif.: Sage.

Neyman, J. 1979. Probability models in medicine and biology: Avenues for their validation for humans in real life. Berkeley: University of California, Statistical Laboratory.

Nisbett, R. E., and L. Ross. 1980. Human Inference: Strategies and Shortcomings of Social Judgment. Englewood Cliffs, N.J.: Prentice-Hall.

Northwest Coalition for Alternatives to Pesticides. 1985. Position Document— Risk Analysis. NCAP News (Winter):33.

Office of Science and Technology Policy. 1984. Chemical carcinogens: Review of the science and its associated principles. Federal Register 49(100):21594–21661.

O'Flaherty, E. J. 1984. Pharmacokinetic methods in risk assessment. In Handbook of Risk Assessment. Washington, D.C.: National Science Foundation.

O'Leary, M. K., W. D. Coplin, H. B. Shapiro, and D. Dean. 1974. The quest for relevance. International Studies Quarterly 18:211–237.

Östberg, G., H. Hoffstedt, G. Holm, B. Klingernstierna, B. Rydnert, V. Samsonowitz, and L. Sjöberg. 1977. Inconceivable Events in Handling Material in Heavy Mechanical Engineering Industry. Stockholm, Sweden: National Defense Research Institute.

Otway, H. J., and D. von Winterfeldt. 1982. Beyond acceptable risk: On the social acceptability of technologies. Policy Sciences 14:247–256.

Page, T. 1978. A generic view of toxic chemicals and similar risks. Ecology Law Quarterly 7:207–243.

Page, T. 1981. A framework for unreasonable risk in the Toxic Substances Control Act. In Carcinogenic Risk Assessment, R. Nicholson, ed. New York: New York Academy of Sciences.

Parducci, A. 1974. Contextual effects: A range-frequency analysis. In Handbook of Perception, Vol. 2, E. C. Carterette and M. P. Friedman, eds. New York: Academic Press.

Payne, S. L. 1952. The Art of Asking Questions. Princeton, N.J.: Princeton University Press.

Pearce, D. W. 1979. Social cost-benefit analysis and nuclear futures. In Energy Risk Management, G. T. Goodman and W. D. Rowe, eds. New York: Academic Press.

Peterson, C. R., and L. R. Beach. 1967. Man as an intuitive statistician. Psychological Bulletin 69(1):29–46.

Peto, R. 1980. Distorting the epidemiology of cancer. Nature 284:297–300.

Pew, R. D., C. Miller, and C. E. Feeher. 1982. Evaluation of Proposed Control Room Improvements Through Analysis of Critical Operator Decisions. Palo Alto, Calif.: Electric Power Research Institute.

Pinder, G. F. 1984. Groundwater contaminant transport modeling. Environmental Science and Technology 18(4):108A–114A.

Poulton, E. C. 1968. The new psychophysics: Six models of magnitude estimation. Psychological Bulletin 69:1–19.

Poulton, E. C. 1977. Quantitative subjective assessments are almost always biased, sometimes completely misleading. British Journal of Psychology 68:409–421.

President's Commission on the Accident at Three Mile Island. 1979. Report of the President's Commission on the Accident at Three Mile Island. Washington, D.C.: U.S. Government Printing Office.

Rayner, S., and R. Cantor. 1987. How fair is safe enough?: The cultural approach to societal technology choice. Risk Analysis 7(1):3–9.

Reissland, J., and V. Harries. 1979. A scale for measuring risks. New Scientist 83:809–811.

Rodricks, J. V., and R. G. Tardiff. 1984. Animal research methods for dose-response assessment. In Handbook of Risk Assessment. Washington, D.C.: National Science Foundation.

Rokeach, M. 1973. The Nature of Human Values. New York: The Free Press.

Roling, G. T., L. W. Pressgrove, E. B. Keefe, and S. B. Raffin. 1977. An appraisal of patients' reactions to "informed consent" for peroral endoscopy. Gastrointestinal Endoscopy 24(2):69–70.

Rosencranz, A., and G. S. Wetstone. 1980. Acid precipitation: National and international responses. Environment 22(5):6–20, 40–41.

Rosenthal, R., and R. L. Rosnow. 1969. Artifact in Behavioral Research. New York: Academic Press.

Rothman, S., and S. R. Lichter. 1987. Elite ideology and risk perception in nuclear energy policy. American Political Science Review 81(2):383–404.

Rothschild, N. M. 1978. Rothschild: An antidote to panic. Nature 276:555.

Rubin, D., and D. Sachs, eds. 1973. Mass Media and the Public. New York: Praeger.

Schnaiburg, A. 1980. The Environment: From Surplus to Scarcity. New York: Oxford University Press.

Schneider, S. H., and L. E. Mesirow. 1976. The Genesis Strategy. New York: Plenum.

Schneiderman, M. A. 1980. The uncertain risks we run: Hazardous material. In Societal Risk Assessment: How Safe is Safe Enough?, R. C. Schwing and W. A. Albers, Jr., eds. New York: Plenum.

Schudson, M. 1978. Discovering the News. New York: Basic Books.

Schwarz, E. D. 1978. The use of a checklist in obtaining informed consent for treatment with medicare. Hospital and Community Psychiatry 29:97–100.

Seligman, M. E. P. 1975. Helplessness. San Francisco: Freeman, Cooper.

Shaklee, H., B. Fischhoff, and L. Furby. 1988. The psychology of contraceptive surprises: Cumulative risk and contraceptive failure. Eugene, Oreg.: Eugene Research Institute.

Sharlin, H. I. 1987. Macro-risks, micro-risks, and the media: The EDB case. In The Social and Cultural Construction of Risk, B. B. Johnson and V. T. Covello, eds. Dordrecht, Holland: D. Reidel.

Sheridan, T. B. 1980. Human error in nuclear power plants. Technology Review 82(4):23–33.

Shroyer, T. 1970. Toward a critical theory for advanced industrial society. In Recent Sociology, Vol. 2, Patterns of Communicative Behavior, H. P. Drietzel, ed. London: Macmillan.

Sioshansi, F. P. 1983. Subjective evaluation using expert judgment: An application. IEEE Transactions on Systems, Man and Cybernetics 13(3):391–397.

Sjöberg, L. 1979. Strength of belief and risk. Policy Sciences 11:539–573.

Slovic, P. 1962. Convergent validation of risk-taking measures. Journal of Abnormal and Social Psychology 65:68–71.

Slovic, P. 1986. Informing and educating the public about risk. Risk Analysis 6(4):403–415.

Slovic, P., and B. Fischhoff. 1977. On the psychology of experimental surprises. Journal of Experimental Psychology: Human Perception and Performance 3:544–551.

Slovic, P., and B. Fischhoff. 1983. How safe is safe enough? Determinants of perceived and acceptable risk. In Too Hot to Handle? Social and Policy Issues in the Management of Radioactive Wastes, C. Walker, L. Gould, and E. Woodhouse, eds. New Haven, Conn.: Yale University Press.

Slovic, P., B. Fischhoff, and S. Lichtenstein. 1978. Accident probabilities and seatbelt usage: A psychological perspective. Accident Analysis and Prevention 17:10–19.

Slovic, P., B. Fischhoff, and S. Lichtenstein. 1979. Rating the risks. Environment 21:14–20, 30, 36–39.

Slovic, P., B. Fischhoff, and S. Lichtenstein. 1980. Facts vs. fears: Understanding perceived risk. In Societal Risk Assessment: How Safe Is Safe Enough?, R. Schwing and W. A. Albers, Jr., eds. New York: Plenum.

Slovic, P., B. Fischhoff, and S. Lichtenstein. 1984. Modeling the societal impact of fatal accidents. Management Science 30:464–474.

Slovic, P., B. Fischhoff, S. Lichtenstein, B. Corrigan, and B. Combs. 1977. Preference for insuring against probable small losses: Implications for the theory and practice of insurance. Journal of Risk and Insurance 44:237–258.

Smith, V. K., and W. H. Desvousges. 1986. Measuring Water Quality Benefits. Boston: Kluwer.

Stallen, P. J. 1980. Risk of science or science of risk? In Society, Technology and Risk Assessment, J. Conrad, ed. London: Academic Press.

Starr, C. 1969. Social benefit versus technological risk. Science 165:1232–1238.

Svenson, O. 1981. Are we all less risky and more skillful than our fellow drivers? Acta Psychologica 47:143–148.

Svenson, O., and B. Fischhoff. 1985. Levels of environmental decisions. Journal of Environmental Psychology 5:55–67.

Thompson, M. 1980. Aesthetics of risk: Culture or context. In Societal Risk Assessment, R. C. Schwing and W. A. Albers, Jr., eds. New York: Plenum.

Tockman, M. S., and A. M. Lilienfeld. 1984. Epidemiological methods in risk assessment. In Handbook of Risk Assessment. Washington, D.C.: National Science Foundation.

Travis, C. C. 1984. Modeling methods for exposure assessment. In Handbook of Risk Assessment. Washington, D.C.: National Science Foundation.

Tribe, L. H. 1972. Policy science: Analysis or ideology? Philosophy and Public Affairs 2:66–110.

Tukey, J. W. 1977. Some thoughts on clinical trials, especially problems of multiplicity. Science 198:679–690.

Tulving, E. 1972. Episodic and semantic memory. In Organization of Memory, E. Tulving and W. Donaldson, eds. New York: Academic Press.

Turner, C. F. 1980. Surveys of subjective phenomena. In The Measurement of Subjective Phenomena, D. Johnston, ed. Washington, D.C.: U.S. Government Printing Office.

Turner, C. F., and E. Martin, eds. 1985. Surveying Subjective Phenomena, Vols. 1 and 2. New York: Russell Sage Foundation.

Tversky, A., and D. Kahneman. 1971. The belief in the "law of small numbers." Psychological Bulletin 76:105–110.

Tversky, A., and D. Kahneman. 1973. Availability: A heuristic for judging frequency and probability. Cognitive Psychology 5:207–232.

Tversky, A., and D. Kahneman. 1974. Judgment under uncertainty: Heuristics and biases. Science 185:1124–1131.

Tversky, A., and D. Kahneman. 1981. The framing of decisions and the psychology of choice. Science 211(4481):453–458.

U.S. Committee on Government Operations. 1978. Teton Dam Disaster. Washington, D.C.: Government Printing Office.

U.S. Government. 1975. Hearings, 94th Cong., 1st Sess. Browns Ferry Nuclear Plant Fire, September 16, 1975. Washington, D.C.: U.S. Government Printing Office.

U.S. Nuclear Regulatory Commission. 1975. Reactor safety study: An assessment of accident risks in U.S. commercial nuclear power plants. WASH 1400 (NUREG-75/014). Washington, D.C.: U.S. Nuclear Regulatory Commission.

U.S. Nuclear Regulatory Commission. 1978. Risk Assessment Review Group to the U.S. Nuclear Regulatory Commission. NUREG/CR-0400. Washington, D.C.: U.S. Nuclear Regulatory Commission.

U.S. Nuclear Regulatory Commission. 1982. Safety Goals for Nuclear Power Plants: A Discussion Paper. NUREG-0880. Washington, D.C.: U.S. Nuclear Regulatory Commission.

U.S. Nuclear Regulatory Commission. 1983. PRA Procedures Guide. NUREG/CR-2300. Washington, D.C.: U.S. Nuclear Regulatory Commission.

Vlek, C. A. J., and P. J. Stallen. 1980. Rational and personal aspects of risk. Acta Psychologica 45:273–300.

Vlek, C. A. J., and P. J. Stallen. 1981. Judging risks and benefits in the small and in the large. Organizational Behavior and Human Performance 28:235–271.

von Winterfeldt, D., R. S. John, and K. Borcherding. 1981. Cognitive components of risk ratings. Risk Analysis 1(4):277–287.

Weaver, S. 1979. The passionate risk debate. The Oregon Journal, April 24.

Weinberg, A. M. 1979. Salvaging the atomic age. The Wilson Quarterly (Summer):88–112.

Weinstein, N. D. 1980a. Seeking reassuring or threatening information about environmental cancer. Journal of Behavioral Medicine 2:125–139.

Weinstein, N. D. 1980b. Unrealistic optimism about future life events. Journal of Personality and Social Psychology 93:806–820.

Weinstein, N. D., ed. 1987. Taking Care. New York: Cambridge University Press.

White, G., ed. 1974. Natural Hazards: Local, National and Global. New York: Oxford University Press.

Wilson, R. 1979. Analyzing the daily risks of life. Technology Review 81(4):40–46.

Wilson, V. L. 1980. Estimating changes in accident statistics due to reporting requirement changes. Journal of Safety Research 12(1):36–42.

Wohlstetter, R. 1962. Pearl Harbor: Warning and Decision. Stanford, Calif.: Stanford University Press.

Woodworth, R. S., and H. Schlosberg. 1954. Experimental Psychology. New York: Henry Holt.

Wortman, P. M. 1975. Evaluation research: A psychological perspective. American Psychologist 30:562–575.

Wynne, B. 1980. Technology, risk and participation. In Society, Technology and Risk Assessment, J. Conrad, ed. London: Academic Press.

Wynne, B. 1983. Institutional mythologies and dual societies in the management of risk. In The Risk Analysis Controversy, H. C. Kunreuther and E. V. Ley, eds. New York: Springer-Verlag.

Zeisel, H. 1980. Lawmaking and public opinion research: The President and Patrick Caddell. American Bar Foundation Research Journal 1:133–139.

Zentner, R. D. 1979. Hazards in the chemical industry. Chemical and Engineering News 57(45):25–27, 30–34.

Appendix D
Availability of Working Papers

Photocopies of the working papers of the Committee on Risk Perception and Communication are available from the National Academy Press, 2101 Constitution Avenue, N.W., Washington, DC 20418.

- Case Study: "The 1980/82 Medfly Controversy in California," by Emory M. Roe.
- Case Study: "Communicating Corporate Disaster: The Aldicarb Oxime Release at the Union Carbide Plant at Institute, West Virginia, on August 11, 1985," by Rob Coppock.

Appendix E
Key Terms and Distinctions

Risk communication practitioners and researchers and the general public often confuse key distinctions such as that between hazard and risk and that between risk communication and risk message. We have therefore categorized terms in order to emphasize such distinctions.

HAZARD   An act or phenomenon posing potential harm to some person(s) or thing(s); the magnitude of the hazard is the amount of harm that might result, including the seriousness and the number of people exposed.

RISK   Adds to the hazard and its magnitude the probability that the potential harm or undesirable consequence will be realized.

* * *

RISK ASSESSMENT   The characterization of potential adverse effects of exposures to hazards; includes estimates of risk and of uncertainties in measurements, analytical techniques, and interpretive models; quantitative risk assessment characterizes the risk in numerical representations.

RISK CONTROL ASSESSMENT Characterization of alternative interventions to reduce or eliminate the hazard and/or unwanted consequences; considers technological feasibility, costs and benefits, and legal requirements or restrictions.

RISK MANAGEMENT The evaluation of alternative risk control actions, selection among them (including doing nothing), and their implementation; the responsible individual or office (risk manager) sometimes oversees preparation of risk assessments, risk control assessments, and risk messages. Risk management may or may not be open to outside individuals or organizations.

* * *

RISK COMMUNICATION An interactive process of exchange of information and opinion among individuals, groups, and institutions; often involves multiple messages about the nature of risk or expressing concerns, opinions, or reactions to risk messages or to legal and institutional arrangements for risk management.

RISK MESSAGE A written, verbal, or visual statement containing information about risk; may or may not include advice about risk reduction behavior; a formal risk message is a structured written, audio, or visual package developed with the express purpose of presenting information about risk.

* * *

RISK COMMUNICATOR/MESSAGE SOURCE The individual or office sending a risk message or interacting with other individuals, groups, or organizations in a risk communication process; may also be the risk manager, risk message preparer, risk analyst, or other expert.

AUDIENCE/RECIPIENTS The recipient(s) of a risk message; almost never a homogeneous group; can include the recipients intended by the preparer of the message as well as others who receive it even though addressed elsewhere.

Index

A

Abortion attitudes, 230
Acceptable risk, 274, 284; *see also* 54–71, 85–90
Access
  to decision-making process, 7, 127–28, 285–86
  to scientific information, 5, 7–8, 114–15, 141–42, 278–80
Accident reports, 255
Accountability, 10, 156–58
Acid rain, 115
Acquired immune deficiency syndrome (AIDS)
  public's knowledge of, 227–28, 290
  risk communication issues, 6, 89, 90, 116, 135–37, 165
  uncertainty of information on, 61, 121
Action threshold, 173
Active public, 101, 102
Administrative Procedures Act of 1946, 16, 73, 100, 125, 128
Advocacy, *see* Influence techniques
Agricultural workers, 32
Agriculture, 59
Agriculture Department, 113
Air bags, 19
Airline accidents, 257
Air pollution, 58, 116
Alcohol information, 17
Alcohol taxation, 19

Aldicarb oxime, 110
Ambiguously worded questions, 228–33, 265
American Bar Association, 178
American Cancer Society, 115
American Chemical Society, 178
American Medical Association, 7, 178
Anchoring, 226
Animal experiments, 39, 40, 58
Appeals to authority, 84–85
Appeals to emotion, 85
Arsenic contamination, 18
Artificial sweeteners, 274
Asarco Corp., 18
Asbestos hazard, 43, 257
Assassinations, 63
Atomic Energy Commission, 120
Attentive public, 101, 102
Attitude surveys, 228–33, 263–66
Audience/Recipients, 322
  audience profiles, 10, 24, 161–62
  "audience/recipients" defined, 322
  characteristics of, 101–2
  concept defined, 271
  effect on message formulation, 282
  proposed consumer's guide, 12–13, 176–79
  psychological principles, 299–304
  relating messages to, 11, 13, 165–70, 181–82
  risk literacy, 13, 182